Separation
and
Purification
Technology

Separation
and
Purification
Technology

edited by

Norman N. Li

Allied-Signal, Inc.
Des Plaines, Illinois

Joseph M. Calo

Brown University
Providence, Rhode Island

CRC Press
Taylor & Francis Group
Boca Raton London New York

CRC Press is an imprint of the
Taylor & Francis Group, an **informa** business

First published 1992 by Marcel Dekker, Inc.

Published 2019 by CRC Press
Taylor & Francis Group
6000 Broken Sound Parkway NW, Suite 300
Boca Raton, FL 33487-2742

© 1992 by Taylor & Francis Group, LLC
CRC Press is an imprint of Taylor & Francis Group, an Informa business

First issued in paperback 2019

No claim to original U.S. Government works

ISBN-13: 978-0-367-45027-4 (pbk)
ISBN-13: 978-0-8247-8721-9 (hbk)

Visit the Taylor & Francis Web site at
http://www.taylorandfrancis.com

and the CRC Press Web site at
http://www.crcpress.com

Library of Congress Cataloging-in-Publication Data

Separation and purification technology / edited by Norman N. Li,
  Joseph M. Calo.
      p.   cm.
    Includes bibliographical references and index.
    ISBN 0-8247-8721-8 (alk. paper)
    1. Separation (Technology).   2. Chemicals--Purification.   I. Li,
  Norman N.   II. Calo, J.M.
  TP156.S45S387   1992
  660'.2842--dc20   92-19323
                                                          CIP

Preface

Separation and purification technology is an intrinsic part of all chemical processing and bioprocessing. It is therefore not surprising that both conventional and novel separation and purification technologies are being continually adapted and developed to address new, important, and difficult problems in such diverse areas as biotechnology, food processing, electronic materials, polymers and ceramics, processing of energy and natural resources, environmental protection, and hazardous waste minimization and remediation. There are mature, established technologies existing side by side with novel, developing, "cutting edge" techniques and approaches. They constitute an area of continual and sometimes rapid evolution which, consequently, is periodically in need of some consolidation and review.

The current volume is directed at addressing some of these needs by chronicling developments and approaches in several selected applications areas. In doing so, it presents a useful blend of recent developments in theory, appplications, process technology, and related economics. Among the comprehensive reviews concerned with new and developing technologies are chapters on the application of membrane-based separations in biotechnology, commercial applications of emulsion liquid membranes, and economic evaluation of membrane technology. Chapters with strong environmental connotations include one on liquid waste concentration by electrodialysis and another on the application of immobilized bioadsorbents for the recovery of dissolved metals. Two other chapters address novel applications of complexation phenomena, while another deals with fundamental work in mass transfer from drops in pulsed sieve-plate extraction columns.

In summary, this volume provides a sampler of recent developments in separation and purification technology.

We wish to thank the staff of Marcel Dekker, Inc., for their editorial assistance and Dr. Jim Zhou for his assistance in preparing the subject index. We would also like to extend our sincere gratitude to the authors of all the chapters, who made this volume possible.

Norman N. Li
Joseph M. Calo

Contents

Preface   iii
Contributors   vii

1 Economic Assessment of Membrane Processes   1

   H. Strathmann

2 Separation of Unsaturates by Complexing with Nonaqueous
   Solutions of Cuprous Salts   19

   George C. Blytas

3 Olefin Recovery and Purification via Silver Complexation   59

   George E. Keller, Arthur E. Marcinkowsky, Surendra K. Verma,
   and Kenneth Dale Williamson

4 Immobilized Bioadsorbents for Dissolved Metals   85

   Charles D. Scott and James N. Petersen

                                                                          v

5   Membrane Separations in the Recovery of Biofuels and
    Biochemicals: An Update Review   99

    Stephen A. Leeper

6   Commercial Applications of Emulsion Liquid Membranes   195

    Robert P. Cahn and Norman N. Li

7   Evaluation of Mass Transfer Coefficients from Single-Drop
    Models in Pulsed Sieve-Plate Extraction Columns   213

    Qian Yu, Weiyang Fei, Lei Xia, and Jiading Wang

8   Liquid Waste Concentration by Electrodialysis   229

    Rémy Audinos

Index   *303*

Contributors

Rémy Audinos   Ecole Nationale Supérieure de Chimie, Institut National Polytechnique, Toulouse, France

George C. Blytas   Corporate Engineering, Shell Development Company, Houston, Texas

Robert P. Cahn   Private Consultant, Millburn, New Jersey

Weiyang Fei   Department of Chemical Engineering, Tsinghua University, Beijing, China

George E. Keller   Technical Center, R&D Department, Union Carbide Chemicals & Plastics Co., Inc., South Charleston, West Virginia

Stephen A. Leeper*   Bioprocess Development, Idaho National Engineering Laboratory, Idaho Falls, Idaho

Norman N. Li   Research and Technology, Allied-Signal, Inc., Des Plaines, Illinois

---

Current affiliation: Clinton Laboratories, Eli Lilly and Company, Clinton, Indiana.

Arthur E. Marcinkowsky   Technical Center, R&D Department, Union Carbide Chemicals & Plastics Co., Inc., South Charleston, West Virginia

James N. Petersen   Chemical Engineering Department, Washington State University, Pullman, Washington

Charles D. Scott   Oak Ridge National Laboratory, Oak Ridge, Tennessee

H. Strathmann   Department of Chemical Engineering, Universität Stuttgart, Stuttgart, Germany

Surendra K. Verma   Technical Center, R&D Department, Union Carbide Chemicals & Plastics Company, South Charleston, West Virginia

Jiading Wang   Department of Chemical Engineering, Tsinghua University, Beijing, China

Kenneth Dale Williamson   Technical Center, R&D Department, Union Carbide Chemicals & Plastics Co., Inc., South Charleston, West Virginia

Lei Xia   Department of Chemical Engineering, Tsinghua University, Beijing, China

Qian Yu   Department of Chemical Engineering, Tsinghua University, Beijing, China

1
Economic Assessment of Membrane Processes

H. Strathmann   *Universität Stuttgart, Stuttgart, Germany*

INTRODUCTION

In recent years membranes and membrane processes have become industrial products of substantial technical and commercial importance. The worldwide sales of synthetic membranes in 1990 were in excess of US 2.0×10^9. Taking into consideration that in most industrial applications membranes account for about 40% of the total investment costs for a complete membrane plant, the total annual sales of the membrane based industry is close to US 5×10^9 [1]. Membranes and membrane processes have found a very broad range of applications. They are used today to produce potable water from seawater, to treat industrial effluents, to recover hydrogen from off-gases, or to fractionate, concentrate, and purify molecular solutions in the chemical and pharmaceutical industry. Membranes are also key elements in artificial kidneys and controlled drug delivery systems. The growing significance of membranes and membrane processes as efficient tools for laboratory and industrial scale mass separations is based on the several properties, characteristic of all membrane separation processes, which make them superior to many conventional mass separation methods. The mass separation by means of membranes is a physical procedure carried out at ambient temperature; thus the constituents to be separated are not exposed to thermal stress or chemical alteration. This is of particular importance for biochemical or microbiological application where often mixtures of sensitive biological materials have to be separated. Furthermore membrane processes are energy-efficient and rather sim-

1

ple to operate in a continuous mode. Up- or downscaling is easy and process costs depend only marginally on the plant size.

In spite of impressive sales and a growth rate of the industry of about 12–15% per year, the use of membranes in industrial scale separation processes is not without technical and economic problems. Technical problems are related to insufficient membrane selectivities, poor transmembrane fluxes, general process operating problems, and lack of application know-how. Economic problems originate from the multitude of different membrane products and processes with very different price structures in a wide range of applications which are distributed on a very heterogeneous market consisting of a multitude of often very small market segments for individual products. This has led to relatively large production volumes for some products, such as hemodialyzers, disposable items used only once for a few hours and sold in relatively large and uniform market segments. Other membranes, such as certain ceramic structures, used in special applications in the food, chemical, or pharmaceutical industry are expected to last for several years in operation and can only be sold in relatively small quantities to small market segments. Consequently, production volumes for these items are low and prices high. The rapid development of new membrane products and processes opening up new applications makes it difficult to predict the growth rate of the market with reasonable accuracy. However, the demand for efficient separation processes in the chemical, food, and drug industry as well as in biotechnology and for solving challenging environmental problems has not only led to an optimistic view but also to a considerable amount of speculation concerning the future development of the membrane based industry.

In this chapter the different membrane products and processes are evaluated in terms of their state of development and their technical and economic relevance. Their potentials and limitations are pointed out. The present market for membranes and related products is analyzed in terms of areas of application, technical requirements, economic limitations, and regional distribution. The structure of today's membrane industry is examined in terms of its product lines, operating strategies, and regional distribution. Main parameters affecting the present and future utilization of membranes and membrane processes, such as membrane performance and costs, process reliability, long-term experience, etc., are described. Different strategies used by various companies for handling their membrane-related business are illustrated and evaluated.

The critical needs in terms of basic and applied research, capital investments, and personal education for a further growth of the membrane industry are discussed. Various research topics are identified as being crucial for the future development of membranes and membrane processes. They are analyzed for their technical relevance, commercial impact, prospects for successful realization, and the financial effort this will require.

FUNDAMENTALS OF MEMBRANES AND MEMBRANE PROCESSES

To better understand the significance, both technical and economic, of membrane processes in solving mass separation problems, some basic aspects concerning membranes and their function shall be briefly reviewed at this point.

Definition of a Membrane: Structure and Function

In a most general sense, a membrane is an interphase separating two homogeneous phases and affecting the transport of different chemical components in a very specific way. A multitude of different structures is summarized under the term *membrane*. Often a membrane can be described easier by the way it functions than by its structure. Three different modes of transport can be distinguished in synthetic membranes, as indicated in Figure 1. Finally, a component may also be

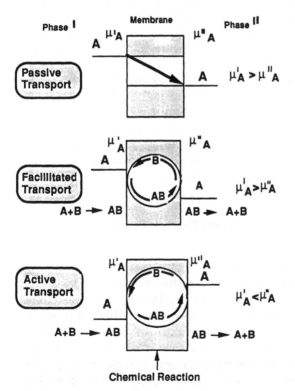

Figure 1   Schematic drawing illustrating the various modes of mass transportation in synthetic membranes.

transported against its chemical potential gradient through the membrane if coupled to a carrier and an energy-delivering reaction. However, this mode of mass transport, referred to as *active transport*, has no up-to-date technical or commercial relevance.

Properties and Applications of Technically Relevant Membranes

Process design and chemical engineering aspects are important for the overall performance of a membrane separation process. The key element, however, is the membrane. Various structures are used as membranes. Some are very simple, such as microporous structures, whereas others are more complex containing functional groups and selective carriers. A summary of technically relevant membranes, their structure, and their area of application is given in Table 1.

The most simple form of a synthetic membrane is the porous plates or foils. These are produced by pressing and sintering a polymeric, a ceramic, or a metal powder. These membranes have relatively large pores, a wide pore size distribution, and a low porosity. The symmetric or asymmetric phase inversion membranes are more complex in their production as well as in their structure. They are produced by precipitation of a polymer solution. These membranes are used today in micro- and ultrafiltration as well as in gas separation and pervaporation. Composite membranes are more and more employed in the last three processes mentioned above. The selective layer and the support structure of these membranes consist of different materials. Membranes with functional groups, like the simple ion exchange membranes, are used in electrolysis and electrodialysis. Liquid membranes that are used in coupled transport with selective complexing agents and chelates are gaining increasing importance.

Technically Relevant Membrane Separation Processes and Their Applications

Membrane processes are just as heterogeneous as the membranes. Significant differences occur in the membranes used, the driving forces for the mass transport, the applications, and also their technical and economical significance. Technically relevant membrane processes and their most important characteristics are given in Table 2. In some of the processes the membranes as well as the processes have reached such a level that completely new developments cannot be expected. Examples of this are the micro- and ultrafiltration, reverse osmosis, dialysis or electrodialysis. In these processes only improvements and optimization of existing systems and their adaptation to special applications are to be expected. Other processes are in the very beginning of their industrial application and offer the possibility of totally new developments of membranes and modules.

Table 1 Properties and Applications of Technically Relevant Synthetic Membranes

Membranes	Basic materials	Manufacturing procedures	Structures	Applications
Ceramic membranes	Clay, silicate, aluminium-oxide, graphite, metal powder	Pressing and sintering of fine powders	Pores of 0.1–10μm diameter	Filtering of suspensions, gas separations, separation of isotopes
Stretched membranes	Polytetrafluoroethylene, polyethylene, polypropylene	Stretching of partially crystalline foil perpendicular to the orientation of crystallyts	Pores of 0.1–1μm diameter	Filtration of aggressive media, cleaning of air, sterile filtration, medical technology
Etched polymer films	Polycarbonate	Radiation of a foil and subsequent acid etching	Pores of 0.5–10μm diameter	Analytic and medical chemistry, sterile filtration
Homogeneous membranes	Silicone rubber, hydrophobic liquids	Extruding of homogeneous foils, formation of liquid films	Homogeneous phase, support possible	Gas separations, carrier-mediated transport
Symmetric microporous membranes	Cellulose derivatives, polyamide, polypropylene	Phase inversion reaction	Pores of 50–5000nm diameter	Sterile filtration, dialysis, membrane distillation
Integral asymmetric membranes	Cellulose derivatives, polyamide, polysulfone, etc.	Phase inversion reaction	Homogeneous polymer or pores of 1–10nm diameter	Ultrafiltration, hyperfiltration, gas separations, pervaporation
Composite asymmetric membranes	Cellulose derivatives, polyamide, polysulfone, polydimethylsiloxane	Application of a film to a microporous membrane	Homogeneous polymer or pores of 1–5nm diameter	Ultrafiltration, hyperfiltration, gas separations, pervaporation
Ion exchange membranes	Polyethylene, polysulfone polyvinylchloride etc.	Foils from ion exchange resins or sulfonation of homogeneous polymers	Matrix with positive or negative charges	Electrodialysis, electrolysis

Table 2 Technically Relevant Membrane Process

Membrane separation process	Driving force for mass transport	Type of membrane employed	Separation mechanism of the membrane	Application
Microfiltration	Hydrostatic pressure difference 50–100 kPa	Symmetric porous membrane with a pore radius of 0.1–20μm	Sieving effect	Separation of suspended materials
Ultrafiltration	Hydrostatic pressure difference 100–1000 kPa	Symmetric porous membrane with a pore radius of 1–20 nm	Sieving effect	Concentration, fractionation and cleaning of macromolecular solutions
Reverse osmosis	Hydrostatic pressure difference 1000–10,000 kPa	Asymmetric membrane from different homogeneous polymers	Solubility and diffusion in the homogeneous polymer matrix	Concentration of components with low molecular weight
Dialysis	Concentration difference	Symmetric porous membrane	Diffusion in a convection-free layer	Separation of components with low molecular weight from macromolecular solutions
Electrodialysis	Difference in electrical potential	Ion exchange membrane	Different charges of the components in solution	Desalting and deacidifying of solutions containing neutral components
Gas separation	Hydrostatic pressure difference 1000–15,000 kPa	Asymmetric membrane from a homogeneous polymer	Solution and diffusion in the homogeneous polymer matrix	Separation of gases and vapors
Pervaporation	Partial pressure difference 0–100 kPa	Asymmetric solubility membrane from a homogeneous polymer	Solution and diffusion in the homogeneous polymer matrix	Separation of solvents and azeotropic mixtures

Membrane Modules and Their Design

For a practical application membranes have to be installed in a suitable device generally referred to as the *membrane module*. The membrane module design is closely related to the membrane process and its technical application. It is performed according to economical considerations, whereby the costs per installed membrane area are an important factor, and to other chemical engineering criteria, such as the flow of the feed solution at the membrane surface and the flow path of the permeate. The goal is to obtain an optimum flow to and from the membrane while minimizing pressure loss, concentration polarization, and membrane fouling. For the different applications, different types of modules have been developed which generally fulfill the given demands. The main types of membrane modules used today on an industrial scale are listed in Table 3.

TECHNICALLY RELEVANT MEMBRANE APPLICATIONS AND THEIR ECONOMICAL SIGNIFICANCE

Membranes are used today in numerous applications, from medicine to wastewater treatment. The applications show great differences in their technical and economical significance.

Table 3 Membrane Modules, Their Properties and Applications

Type of module	Membrane area per volume (m²/m³)	Price	Control of concentration polarization	Application
Pleated filter cartridge	800–1000	Low	Very poor	Dead-end filtration
Tube	20–30	Very high	Very good	Cross-flow filtration of solutions with high solids content
Plate-and-frame	400–600	High	Fair	Filtration, pervaporation, gas separation, reverse osmosis
Spiral-wound	800–1000	Low	Poor	Ultrafiltration, reverse osmosis, pervaporation, gas separation
Capillary tube	600–1200	Low	Good	Ultrafiltration, pervaporation, liquid membranes
Hollow fiber		Very low	Very bad	Reverse osmosis, gas separation

Membrane Processes in Water Treatment

Water treatment is one of the most important application areas of membranes. This includes the production of potable water from sea- and brackish water as well as the production of process water and the cleaning of industrial wastewater streams. Micro- and ultrafiltration as well as reverse osmosis and electrodialysis are the processes employed mainly in water treatment. Table 4 shows the segmentation of the total market for membranes and membrane modules, respectively, in relation to the different processes. The table shows that total sales of the membrane industry in the area of water treatment is US $500 million per year. Of all membrane processes, microfiltration plays the most important economical role because of its application in the production of ultrapure water in the semiconductor industry and of sterile and pyrogen-depleated water for the chemical and pharmaceutical industry. The main application of reverse osmosis is in the desalination of sea- and brackish water, but more recently reverse osmosis is also finding increasing use in the production of ultrapure water. The level that has been reached in reverse osmosis membrane performance today makes it unlikely that membranes with revolutionary improved properties will be available in the near future. This, however, does not mean that further development is not necessary. The weak point of all available reverse osmosis membranes today is their poor thermal and chemical stability. All commercial membranes are either sensitive against hydrolysis or not stable in the presence of free chlorine or other strong oxidizing agents. This makes a rather extensive pretreatment of the raw water necessary. In ultrafiltration membrane fouling is a problem that has not yet been solved.

Table 4 Sales of Membrane Industry (million US $ per year) in Water Treatment, Divided by Processes and Applications

Applications	Micro-filtration	Ultra-filtration	Reverse osmosis	Electro-dialysis	Total
Desalting of seawater	—	—	20	—	20
Desalting of brackish water	—	—	35	35	70
Pretreatment of boiler and feed-water	5	—	25	10	35
Ultrapure water	220	30	15	—	265
Sterile and low-pyrogen water	80	10	10	—	100
Industrial wastewater	5	20	15	15	55
Total	310	60	120	60	490

Table 5 Sales of Membrane Industry (million US $ per year) in the Food Industry, Divided by Processes and Applications

Applications	Microfiltration	Ultrafiltration	Reverse osmosis	Electrodialysis	Total
Dairy industry	—	35	—	10	45
Beverage industry	95	5	5	5	110
Meat processing	—	2	5	—	7
Starch industry	—	2	5	—	7
Total	95	44	15	15	169

Membrane Processes in the Food Industry

Another interesting area of application for membranes is the food industry, as indicated in Table 5, which shows the total annual sales of membranes and membrane modules in the food industry. Micro- and ultrafiltration, reverse osmosis, and electrodialysis are mainly used in this application. Microfiltration dominates in the beverage industry for sterile filtration and clarification, while ultrafiltration and electrodialysis are used mainly in the dairy industry. The application of new membrane processes like pervaporation is just beginning. These processes will soon be used on an industrial scale. The overall membrane market in this area amounts to about US $170 million per year.

Membrane Processes in Medical Devices

Membranes have been used in medical applications for more than 25 years. The artificial kidney is by far the most interesting membrane application besides membrane-controlled therapeutic systems for the controlled release of drugs. Plasmapheresis also is now gaining importance, as shown in Table 6, which lists the annual sales of membranes in medical applications.

Membrane Processes in the Chemical and Petrochemical Industry

The use of membrane processes in the chemical and petrochemical industry is rapidly increasing. However, the sales in this area are comparatively small today, in spite of a large potential market. This is shown in Table 7, which is a summary of the annual sales of membranes and membrane modules in the chemical and petrochemical industry.

Table 6 Sales of Membrane Industry (million US $ per year) in Medical Devices, Divided by Processes and Applications

Applications	Microfiltration	Ultrafiltration	Dialysis	Total
Hemodialysis	—	—	900	900
Hemofiltration	—	120	—	120
Plasmapheresis	20	10	—	30
Total	20	130	900	1050

An interesting application is the separation of gases, such as the recovery of hydrogen from the ammonia synthesis and the separation of carbon dioxide from methane in enhanced oil recovery. The separation of oxygen and nitrogen by membranes is of interest for the production of oxygen-enriched air or inert gas. But membranes have to be improved in terms of selectivity and permeability before oxygen-nitrogen separation on a large scale becomes economically feasible.

Pervaporation and perstraction are relatively new membrane processes with numerous applications in the chemical and petrochemical industry.

Membrane Processes in Biotechnology

Biotechnology is an industry where membranes offer great potential advantages. They are especially suitable for the separation of sensitive biological substances because the separation by membranes is a physical procedure which can be carried out at room temperature. All membrane processes including electrodialysis and pervaporation are of interest in biotechnology. Presently, however, sales of membranes for biotechnology applications are still rather low and in the order of US $10–20 million. A reason for the relatively slow introduction of membrane processes in biotechnology is that the modern industrial biotechnology is still very young, and membranes and membrane processes have not yet been sufficiently adapted to the specific requirements of biotechnology. The list of industrial applications of membranes which is given here is far from complete. There are many other applications, especially in the laboratory, which contribute substantially to the total sales of the membrane industry.

THE MEMBRANE MARKET AND ITS EXPECTED DEVELOPMENT

The worldwide sales of all membranes and membrane modules in 1990 were in excess of US $2000 million. There are, however, large differences as far as the different products and their regional distribution is concerned.

Table 7　Sales of Membrane Industry (million US $ per year) in Chemical Industry, Divided by Processes and Applications

Applications	Microfiltration	Ultrafiltration	Reverse osmosis	Gas separation	Pervaporation	Electrodialysis	Electrolysis	Total
Process water pretreatment	25	—	5	—	—	—	—	30
Fractionation of molecular mixtures	10	15	5	—	—	—	—	30
H_2 recycling	—	—	—	20	—	—	—	20
N_2 production	—	—	—	15	—	—	—	15
CO_2/CH_4 separation	—	—	—	5	—	—	—	5
O_2 enrichment	—	—	—	5	—	—	—	5
Separation of azeotropic mixtures	—	—	—	—	5	—	—	5
Desalting of process solutions, salt production	—	—	—	—	—	20	—	20
Chlorine alkaline electrolysis	—	—	—	—	—	—	70	70
Total	35	15	10	45	5	20	70	200

Process and Product Related Market Distribution

The distribution of the world membrane market according to processes is given in Figure 2. Here annual sales of membranes in the various processes and the expected growth of the market during the next 5 years are shown. The largest market for membranes today is that of dialyzers used in artificial kidneys. Sales in other areas, such as ultrafiltration, electrodialysis, and, in particular, gas separation and pervaporation, are presently of minor importance. The expected increase in sales in these processes is, however, significantly higher than in dialysis or microfiltration. While the annual increase in sales of all membrane products is expected to be in the order of 12–15%, the sales increase will be significantly lower than that of the more recently developed processes, such as gas separation or pervaporation.

Regional Distribution of the Membrane Market

The membrane market is rather unevenly distributed, with about 75% of the market being located in the industrialized countries of Europe, USA, and Japan, as indicated in Figure 3, which shows the regional distribution of the membrane market [2]. This is, however, a very global analysis. The distribution of the market can be totally different for the different membrane processes or membrane applications. For example, the market share of Germany in the area of the artificial kidney is 13% and that for electrodialysis only 1% of the entire world market.

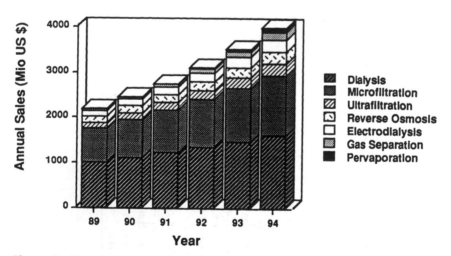

Figure 2 Present and expected future annual sales of membranes according to the different processes.

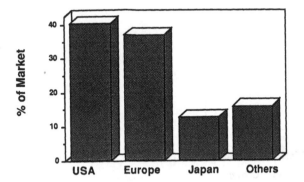

Figure 3 Regional distribution of the membrane market.

It is also obvious that the existence of a regional market is of great significance for the buildup of the related industry. Therefore, there is no manufacturer of ion exchange membranes in Germany, but almost 80% of all hemodialysis membranes are manufactured here.

On the other hand, electrodialysis is used mainly for brackish water desalination and in Japan for the production of bath salt. Both applications are of little significance in Europe; consequently, there are no ion exchange membrane manufacturers in Europe.

Regional Distribution of the Membrane Industry

A look at the regional distribution of the membrane industry provides some interesting information. Almost all companies operate internationally and market their products worldwide. Worldwide more than 100 companies are involved in one way or another in membrane technology [3]. However, only about 60 companies are at the same time manufacturers of membranes or modules on a commercial basis. The other companies are mainly involved in process design and plant engineering using membranes as components. The USA has a large part of the membrane-based industry with more than 35 companies offering membranes on a commercial basis. However, six of these companies account for 80% of the sales of all the USA-based membrane industry. An analysis of the regional distribution of the sales in the membrane industry in relation to the location of the companies shows that companies based in the USA account for 63% of the total worldwide sales of about US $2 billion per year. Japanese companies account for 17% and companies based in European countries for 20% of the worldwide membrane sales. This regional sales distribution is shown in Figure 4.

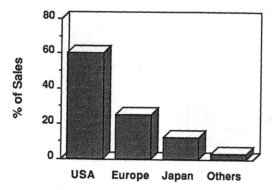

Figure 4 Regional distribution of the membrane industry.

Price Structure of Membrane Products and Processes

The prices of membrane products are closely related to their sales volume. For instance, the sales price for an ultrafiltration-type membrane in an artificial kidney including the module is in the order of US $20 per m^2 membrane area. The same membrane in an industrial ultrafiltration process will probably be in the order of US $200 per m^2. A membrane used in pervaporation or gas separation, in principle not significantly more difficult to produce, will cost in excess of US $500. The difference can be found in process application. To produce 1 m^3 potable water from brackish water by reverse osmosis or electrodialysis will cost US $1–3.

To treat 1 m^3 of wastewater or a process stream by reverse osmosis or ultrafiltration usually will be significantly higher and can easily reach US $10–20, depending on the feed solution constituent. Again here the process costs are highly affected by the market size.

THE MEMBRANE INDUSTRY AND ITS STRUCTURE

The application of membranes is extremely widespread involving a multitude of different structures and processes. Similarly heterogeneous is the membrane industry. Being successful in the market requires, in addition to an appropriate membrane as the key element, a multitude of peripheral components, which consist of special hardware, such as pumps or monitoring equipment, and software, such as basic engineering, process design, and especially application know-how.

The complexity of the membrane-based industry is indicated in the schematic diagram of Figure 5, which shows on the horizontal level the various technologies required to penetrate the market. These technologies are membrane research and development, membrane production, basic engineering, process design, and

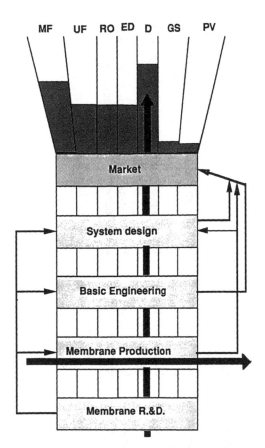

Figure 5 Schematic diagram illustrating the structure of the membrane industry. MF = microfiltration, UF = ultrafiltration, RO = reverse osmosis, D = dialysis, GS = gas separation, PV = pervaporation.

marketing. In the vertical direction the various membrane processes are indicated with fractional sales, which are highest in dialysis and microfiltration and lowest in pervaporation. Since every membrane process requires the various levels of technology, it is difficult for one company to cover all membrane processes on all levels of technologies and applications.

Companies have often specialized in certain areas. Some only produce membranes for various processes or applications and sell their products through various equipment manufacturers. This is indicated by the horizontal arrow in Figure 5. Other companies concentrate on one process, e.g., producing dialyzers for the artificial kidney. This is indicated by the vertical arrow in Figure 5. These companies usually are relatively large, doing their own research, basic engineer-

Table 8 State of Development of Membrane Processes with Industrial Use and Present R&D Efforts to Improve Membranes and Processes in New Applications

Process	Availability	Problems	Membranes	Goals of R&D work	Applications
Microfiltration	State of the art	Membrane life	Asymmetric, inorganic	Cross-flow filtration	Biotechnology
Ultrafiltration	State of the art	Membrane fouling	Surface modifications	Membrane cleaning	Biotechnology, chemical industry
Reverse osmosis	State of the art	Chlorine stability, biofouling	Chlorine resistance	Water pretreatment	Biotechnology, wastewater
Dialysis	State of the art	Biocompatibility	Improved permeability	Process optimization	Biotechnology, food technology
Electrodialysis	State of the art	Alkaline stability	Ionselective and bipolar membranes	Multiple cell systems	Chemical industry, wastewater
Gas separation	Pilot plants	Membrane selectivity and flux	Membranes for specific gases	Process optimization	Air/solvent separation
Pervaporation	Laboratory loops	Membrane selectivity, process development	Membranes for specific materials	Process development and optimization	Wastewater treatment, biotechnology

ing, process design, and marketing. Both business approaches seem feasible as very successful companies have shown.

MEMBRANE-RELATED RESEARCH AND DEVELOPMENT

The R&D work in membrane science and technology is carried out at different levels and with different goals. A complete evaluation of the R&D activities is extremely difficult to carry out. All data concerning this topic are therefore questionable and can only be considered as an analysis of current trends. It is often difficult and arbitrary to decide which activities are R&D "product development," or quality control.

The R&D activities carried out in the industry often differ significantly from such activities carried out at universities in terms of their goals. The research in industry is in general dictated by economical considerations and concentrated on the development and optimization of products and processes. The universities are more involved in basic research which is at least to some extent independent of economical considerations. Since the various membrane processes show great differences in their level of technical development, research work is concentrated on different topics in the different processes.

Table 8 shows some of the subjects on which R&D effects are concentrated in membrane processes available today. Processes which are state-of-the-art and used industrially include microfiltration, ultrafiltration, reverse osmosis, dialysis, electrodialysis, and, with some restrictions, pervaporation and gas separation.

These processes have the greatest economic relevance today. The R&D work in this area is concentrated on product improvement as well as process optimization and different aspects of application. Gas separation and pervaporation are of minor economical importance today, and the membrane sales for these processes are hampered by lack of process experience. There is a number of membranes and membrane processes which exist today only on laboratory scale or as concepts, such as membranes with active transport properties, membranes in energy generation or energy conversion, etc. It seems too early to speculate about the future economic significance of these membranes.

REFERENCES

1. R. W. Baker, et al., Membrane Separation Systems—A Research and Development Needs Assessment, National Technical Information Service, U.S. Department of Commerce, NTIS-PR 360, 1990.
2. Proceedings of Aachener Membran Kolloquium, Aachen, March 16–18, 1987, GVC VDI Gesellschaft für Verfahrenstechnik, Düsseldorf.
3. H. Strathmann, Economic evaluation of the membrane technology. In *Future Industrial Prospects of Membrane Processes*, (L. Cecille and J.-C. Toussaint, eds.), Elsevier, Amsterdam, 1989.

2

Separation of Unsaturates by Complexing with Nonaqueous Solutions of Cuprous Salts

George C. Blytas   *Shell Development Company, Houston, Texas*

INTRODUCTION

Since the platinum(II)-chloride-ethylene complex was first discussed by Zeise in 1830 [1] olefin complexes of Ag(I), Cu(I), and Hg(II) have been extensively studied [2]. In 1951, Dewar [3] discussed Ag(I)-olefin complexes using molecular orbital treatment. In 1953, Chatt and Duncanson [4] extended the treatment to Pt(II)-olefin complexes, and in 1973 Solomon and Kochi [5] extended the treatment to Cu(I)-olefin complexes.

Metal-olefin complexes have elicited considerable interest for their potential use in chemical and catalytic reactions. In addition, since olefin complexation in Ag(I) and Cu(I) systems is reversible, these metals have also received attention as potential separating agents for olefin/paraffin separations. The first commercially successful process for separation of unsaturates via Cu(I) complexation is the cuprammonium acetate process, developed during World War II, for the recovery and purification of butadiene used in synthetic rubber [6]. However, the cuprous salt–aqueous solvent system used in this process cannot be used to complex olefins or diolefins other than ethylene and butadiene. The use of solid cuprous halides for the separation of light olefins and diolefins with up to four carbons has also been discussed [7]. Again, higher olefins do not form complexes with solid, covalently bonded cuprous halides.

This chapter discusses how the chemical environment of the cuprous ion can be tailored through proper selection of the anion and the solvent to allow complexation and separation of olefins with up to 16 carbon atoms.

THE CUPROUS ION–OLEFIN BOND IN SEPARATIONS

In general, the molecular orbital treatment of the metal-olefin interaction describes the bond in terms of σ bond from the olefin to the metal and a π bond from the metal to the olefin. The σ bond is due to overlap of π electrons of the olefin with a vacant σ orbital of the metal atom. The π backbonding is formed by overlap of the filled d-orbital of the metal with the vacant π^* backbonding orbitals of the olefin. In such complexes low molecular weight olefins and diolefins, in which the double bond is not shielded by bulky alkyl groups, form more stable bonds than high molecular weight olefins or olefins with shielded double bonds [8,9]. Thus, the aqueous cuprammonium acetate system, for example, is not effective in complexing isoprene, which has one more methyl group than the butadiene molecule. The inability of cuprous acetate (in aqueous ammonia) to complex isoprene can be explained in molecular terms as a steric or shielding effect, or in thermodynamic terms as an entropic effect.

An approach which has been proven successful in overcoming those entropic and steric impediments to complexation is the use of cuprous salts of strong acids, preferably with anions rich in oxygen or halogen atoms. Such anions do not form bridging ligands, such as the halide ions in cuprous halide salts. We call the cuprous ion in such salt systems ionic cuprous.

To obtain further insight into the complexing potential of ionic cuprous, consider a system consisting of a cuprous salt, a solvent, an olefin, and possibly other ligands. The competitive equilibria involved in such a system are shown below. Here Un stands for the olefin unsaturate and L may represent any other ligand present such as an anion, a solvent molecule, or other ligand.

$$2Cu(I) \rightleftharpoons Cu^0 + Cu(II) \tag{1}$$

$$Cu(II) + nL \rightleftharpoons Cu(II)Ln \tag{2}$$

$$Cu(I) + mL \rightleftharpoons Cu(I)Lm \tag{3}$$

$$Cu(I) + Un \rightleftharpoons Cu(I)Un \tag{4}$$

$$Cu(I)Lm + Un \rightleftharpoons Cu(I)LmUn \text{ or } Cu(I)Lm\text{-}Un \tag{5}$$

COMPETITIVE EQUILIBRIA IN THE COMPLEXATION OF CUPROUS SALTS

Equation (1) shows the disproportionation reaction which is driven to the right in aqueous systems by the greater stability of the aquo-cupric in comparison with the aquo-cuprous complex [10]. Water and other ligands classifiable as hard bases [11] coordinate Cu(II) more effectively than Cu(I). Coordination of Cu(II) drives reaction (2) to the right and results in eventual depletion of Cu(I) species. For example, the equilibrium constant for reaction (1) in water at 25°C is 10^6.

$$K_{H_2O}^{25°C} = \frac{[Cu(II)]}{[Cu(I)]^2} = 10^6 \tag{6}$$

To avoid extensive disproportionation in aqueous systems, it is necessary to overcome the effect of reaction (2) by the competing reaction (3). This can be accomplished by introducing ligands, such as ammonia, which favor monovalent copper. Thus, the disproportionation equilibrium constant of Cu(I) in aqueous ammonia at 25°C is 10^{-2}.

The addition of stabilizing ligands to aqueous solutions makes it possible to increase the concentration of cuprous copper, but only at the expense of lowering the activity of the uncomplexed ion which is needed to interact with the olefin double bonds. The presence of water in a system, in addition to dictating the use of competing ligands, increases the activity coefficients of high molecular weight hydrocarbons [12]. The resulting decrease in solubility of hydrocarbons in aqueous systems tends to offset the affinity of olefins for the cuprous ion.

These unfavorable factors can be overcome by using a nonaqueous solvent to minimize disporportionation, by avoiding stabilizing ligands to minimize the competition depicted in (3), and by using an ionic cuprous salt so that a high concentration of "free" cuprous ion can be obtained to drive olefin complexation (4) to completion. Of course, in the presence of ligands which complex cuprous ion, mixed ligand-olefin complexes via reaction (5) are possible. An additional complication in cuprous systems is the ready oxidation to cupric ion. Use of an inert atmosphere and stable solvents surmounts this problem.

The first significant work with essentially noncomplexed ionic cuprous ion was that of I. V. Nelson and coworkers [10] in 1961. They showed polarographically that the cuprous state can be stable, even in noncomplexing solvents, provided moisture is rigorously excluded. At about the same time, McCaulay reported on the aromatic complexes of cuprous tetrafluoroborate [13], and Turner and Amma reported on the aromatic complexes of cuprous tetrachloroalanate [14], both cuprous salts of strong acids.

Following those leads and using the model described in Eqs. (1)–(5), we found that ionic cuprous solutions could be made and used in olefin separations. Furthermore, the region of stability of cuprous solutions was found to extend from the dilute solutions (under 1% by weight of salt) tested polarographically by Nelson [10] to ~65 wt % of cuprous salt prepared by this author.

The earliest disclosure on the use of ionic cuprous salt/nonaqueous solvent systems in separations of olefins was made in 1968 [15] and 1970 [16]. In these references, cuprous trifluoroacetate (CuTFA) was shown to form complexes with detergent range olefins when dissolved in aromatic solvents or propionitrile. Other cuprous salts studied in separations of olefins include the nitrate [17,18] and the sulfate [19,20] in propionitrile, and the chloralanate in aromatic solvents [21,25].

PREPARATION OF IONIC CUPROUS SALTS

A large number of reactions can be used to prepare ionic cuprous salts, dry or in solution. The preparations discussed here have been checked by standard elemental analysis of the produced salts after appropriate purification. Yields in excess of 80% have been obtained in most preparations.

Dry Cuprous Salts

Most of the preparations in this class make use of aromatic solvents from which cuprous salts can be recovered easily.

Addition of Anhydride and Cu_2O

The addition of acid anhydride to Cu_2O in an appropriate solvent is an attractive method for the preparation of a cuprous salt in the anhydrous form. The reaction makes full use of the reactants and does not depend on the availability of a precursor salt. This method was used in the preparation of CuTFA, which in turn has been used as a precursor for the preparation of several other cuprous salts.

The reaction

$$Cu_2O + (CF_3CO)_2CO \xrightarrow[HCOOCF_3]{aromatic} 2CuOOCCF_3 \tag{7}$$

is completed by refluxing in benzene solvent for 1–2 hrs, under nitrogen. Dry CuTFA can be recovered from the supernatant solution by evaporation of the solvent or by precipitation of the cuprous salt in excess paraffin followed by drying. The second method is more rapid but gives lower yields.

Reaction of the Acid of the Salt with CuTFA

Any cuprous salt, CuA, whose acid HA boils above the boiling point of trifluoro-acetic acid (70°C) can be prepared by the reaction:

$$CuOOCCF_3 + HA \xrightarrow[(CF_3CO)_2O]{aromatic} CuA + HOOCCF_3 \tag{8}$$

Good yields and high purity have been achieved using this technique. Since many acids boil above 70°C, this method has wide applicability. On the other hand, this technique is predicated on the availability of CuTFA. Most cuprous salts are less soluble in the aromatic solvent than CuTFA, so that the product usually precipitates as a solid. Best results are obtained when the acid HA is soluble in the aromatic solvent.

The following salts were prepared in the dry form using this technique: cyprous salicylate, cuprous ditertiarybutyl salicylate (CuDTBS), cuprous difluorophosphate, and $CuOSO_2X$ where X is fluorine, methyl, ethyl, or phenyl group.

Reaction Between a Cuprous Halide and Group III Halides

The preparation of $CuAlCl_4$ and $CuAlBr_4$ by fusion of the cuprous and aluminum chlorides have been known for some time [29]. The preparation of dry $CuAlCl_4$ by reacting the two halides in benzene followed by drying has been described [14]. The crystal structure of the $CuAlCl_4$-benzene complex has also been described in the literature. Dry $CuBF_4$ cannot be obtained by this reaction because neither CuF nor $CuBF_4$ is stable in the dry form.

Cuprous Salts in Nonaqueous Solutions

Preparations of solutions of cuprous salts in nonaqueous solvents can be accomplished by redox reactions or by double displacement.

Reduction of Cupric Salts by Metallic Copper in Nitriles

The reduction of anhydrous cupric salt by metallic copper in refluxing propionitrile (EtCN) has been used for the preparation of CuTFA/propionitrile, Cu_2SO_4/propionitrile, and $CuNO_3$/propionitrile.

$$CuSO_4 + Cu^0 \xrightarrow{EtCN} Cu_2SO_4 \qquad (9)$$

The driving force in reaction is the stabilization of the monovalent copper by the nitrile group. The reaction depends on the availability of the anhydrous cupric salt.

Propionitrile is an ideal solvent for this reaction because most cuprous salts are very soluble in it. On the other hand, the affinity of nitriles for cuprous salts makes the recovery of the dry cuprous salts from this solvent difficult. From a practical standpoint, complete removal is not necessary when the propionitrile is used as the solvent in a separation process. Under these conditions, partial removal of the nitrile is sufficient.

Reductive Displacement of Silver by Metallic Copper

The reaction between a silver salt and metallic copper can be used to prepare cuprous salt in any solvent in which the cuprous salt is stabilized by solvation.

$$AgA + Cu^0 \xrightarrow{solvent} CuA + Ag^0 \qquad (10)$$

The salt $CuBF_4$ has been prepared by reaction (10) in benzene and nitrile, and $CuPF_6$ was prepared in an aromatic solvent.

Double Decomposition

By taking advantage of the high solubility of cuprous salts in popionitrile and of the low solubility of AgCl in the same solvent, we can prepare the cuprous salt from its silver counterpart by double displacement. The method is limited to Ag salts which have some solubility in nitriles.

$$CuCl + AgA \xrightarrow{\text{EtCN}} AgCl \downarrow + CuA \tag{11}$$

This method was used for the preparation of $CuOOCC_2F_5$ and Cu_2SO_4. Nearly quantitative removal of the nitrile from the perfluoropropionate was accomplished by low-pressure evaporation followed by azeotroping with n-heptane and benzene. Complete removal of propionitrile from the sulfate results in decomposition of the Cu_2SO_4.

SOLUBILITIES OF CUPROUS SALTS IN NONAQUEOUS SOLVENTS

The energetics of solubilization of a salt in a given solvent are controlled by (a) the energy of coordination, (b) the polarity of the salt (ionicity) and solvent (dielectric constant), and (c) the crystal energy of the salt or its solvates.

In the solubilization of cuprous salts, the coordination energy is the most important factor. Experimental solubilities of various cuprous salts in nonaqueous solvents are shown in Table 1. The Cu(I) and Cu(II) concentrations have been determined as discussed in Appendix I.

Most of the cuprous salts are very soluble in coordinating solvents such as nitriles, amines, alkyl sulfides, and alkyl thiocyanates. The cuprous ion content of these solutions at ambient temperatures ranges from 6 to 20 wt % and in the case of CuCl it reaches 40 wt %. The only salts which are not extremely soluble in nitriles and amines are the methylsulfonate, fluorosulfonate, and fluoroborate, which form solid solvates.

Many of the cuprous salts are soluble in olefins and diolefins, which are also coordinating solvents. Such salts are CuTFA, $CuPO_2F_2$, $CuAlCl_4$, and CuDTBS. The fact that the relatively polar salts can be dissolved in the nonpolar olefins indicates that the π-olefin bonds are comparable in strength to those resulting from nitrile or amine coordination. This conclusion is supported by results of thermometric titrations, discussed below.

The solubility of cuprous salts in aromatics is generally low, but there are important exceptions. The CuTFA is soluble by virtue of both low crystal energy and coordination. The CuDTBS dissolves because the large *tert*-butyl groups on the aromatic ring render it hydrocarbonlike. The interaction between the cuprous ion and the aromatic is not a sufficient driving force to solubilize cuprous salicylate. The $CuAlCl_4$ and $CuBF_4$ are dissolved through coordination. Such CuX:Lewis acid salts form stable complexes with aromatics. Generally, however, the coordinating strength of the aromatic is not sufficient to dissolve the cuprous salt. Thus, although $CuPO_2F_2$ dissolves in olefin (through coordination), it does not dissolve in aromatics.

Table 1 Solubilities of Cuprous Salts in Nonaqueous Solvents at 25°C Expressed at wt % of Cuprous Ion in Solution

Anion	ΦH	Pentene-1	Isoprene	MeCN	EtCN	ProNH$_2$	EtSCN	ProCO$_3$	Melting point °C
CF_3COO^-	13	>14.0		12.0	16.0	18.0		13.4	145
$C_2F_5COO^-$	2.5	12.6	17.0		8.8	13.95		3.0	
CH_3COO^-					8.8				
SO_4^{2-}	<1				12.0				Unstable
NO_3^-	<1				18.5				Unstable
Cl^-	<1	2.6		6.5	40.0				422
$PO_2F_2^-$	<1	>15.0	Polymerizes		12.1	3.8(s)	5(s)	Insol.	237
OH–⟨C$_6$H$_4$⟩–COO⁻	<1	>2.0			3.7	16.9	15.8	1.0	250
OH–⟨C$_6$H$_3$⟩–COO⁻	10				13.0	9.2	8.8	2.0	250
⟨C$_6$H$_5$⟩–SO$_3^-$	<2			12.9	10.0	16.6			250
$EtSO_3^-$	<1			6.8	4.8	12.5			150
$MeSO_3^-$	<1				a	a			240 d
FSO_3^-					a	6(s)			N.D.
$AlCl_4^-$	13.5	17.0	Polymerizes	12.0	12.0	Decomposes		Insol.	233

aInsoluble solvates of composition $CuSO_3CH_3\cdot4CH_3CN$ and $CuSO_3CH_3\cdot4CH_3CN$ and $CuSO_3CH_3\cdot2CH_3CN$ were found.

HYDROCARBON CAPACITY OF CUPROUS SALT SOLUTIONS

The solubility of unsaturated and saturated hydrocarbons in nonaqueous cuprous salt solutions is a result of the complexing strength of the cuprous salt and the polarity of the resulting solution. High complexing strength favors high solubility of unsaturates, while high polarity (or salting out) results in low hydrocarbon capacity. In general, nonaqueous solutions of cuprous salts show hydrocarbon capacities higher than those exhibited by aqueous cuprous systems. Solubilities of C_5 hydrocarbons in solutions of cuprous salts in benzene, propionitrile, and propylamine are given in Table 2.

Table 2 suggests that cuprous salts of organic acids (CuTFA, CuDTBS) dissolved in aromatic solvents cannot be used for solvent extraction of olefins because the solubilities of saturated hydrocarbons in such systems are too high. In EtCN solutions solubilities of hydrocarbons are lower and they follow the order diolefin > monoolefin > paraffin, suggesting that solvent extraction in such systems is feasible.

CUPROUS-OLEFIN COMPLEXING WITH IONIC CUPROUS SALTS

In discussing the concept of ionic cuprous salts the thesis was made that such salts would form stronger olefin complexes than more covalent cuprous salts (such as CuCl). Evidence supporting this thesis has been obtained from several experimental observations. Two sets of data will be discussed. The first set pertains to the olefin-complexing ability of CuTFA in the solid state and in solution. These results were obtained by vapor–liquid equilibrium determinations. The second set of data is a calorimetric screening of the olefin-complexing ability of several cuprous salts in toluene, propionitrile, and propylamine. These calorimetric determinations were made with the object of evaluating the effect of the solvent and of the anion coordination on olefin complexing. Both sets of data show that ionic cuprous salts complex olefins containing more than four carbons and that the resulting complexes are stronger than those formed by covalent cuprous salts.

Complexes Formed by Solid CuTFA

The most direct comparison of complexing ability of cuprous salts is that based on the stability of solid cuprous salt–olefin complexes. In such a comparison, solvent effects are excluded and only anion coordination and crystal energy effects are involved.

The dissociation constants of pentene-1 and benzene complexes of CuTFA were determined by measuring the vapor pressure of the hydrocarbon in equilibrium with the solid complex in a McBain balance. The results of these determinations are given in Table 3. This table shows that the stability of the olefin-CuTFA

Table 2 Solubilities of C_5 Diolefins, Monoolefins and Paraffins in Saturated Cuprous Salt Solutions in Nonaqueous Solvents at 25°C

Solvent	Salt	Wt % Cu(I) in solvent, hydrocarbon-free basis	Hydrocarbon solubility in moles of hydrocarbon/mole of Cu(I)			
			Isoprene[a]	Pentene-1	2-Me-Butene-2	Pentane
Benzene	CuTFA	13	b	∞		∞
	CuDBTS	10		∞		∞
	CuAlCl₄	13.5	c	∞		~0.50
Propionitrile	CuTFA	14	∞	3.3	1.64	0.19
	CuTFA	11[d]	∞		3.20	
	Cu₂SO₄	18	0.4		0.19	
	Cu₂SO₄	11[d]	2.3		0.77	
	CuAc	9	1.7		0.38	
	CuNO₃	8.5	4.2			
	CuNO₃	8.5	4.2			
	CuPO₂F₂	12	9.0	2.6		0.40
	CuDTBS	14[d]		>8.0		~0.10
	CuSO₃Φ	10	~0.5			0.20
	CuAlCl₄	12	~0.8			0.036
	CuCl	40		0.17		0.36
	CuCl	15[d]		1.33		0.80
Propylamine	CuTFA	18	13.8	6.5		
	CuSO₃Φ	16.6	7.7	3.0		
	CuEt_Φ	12.5	25.0	7.2		
	Cu salic.	16.6	22.0	5.5		

[a]The ratio c = c/Cu(I) in isoprene solubilities is twice the ratio of hydrocarbon/Cu(I).
[b]Diolefins form insoluble complexes in CuTFA/aromatics.
[c]Isoprene polymerizes in CuAlCl₄ or CuAlCl₄/aromatics.
[d]Unsaturated solutions. Saturated CuDTBS/EtCN contains 17.7 wt % Cu(I).

Table 3 Stability of Cuprous Trifluoroacetate
Complexes as Determined by McBain Balance

Complex	Temp. (°C)	Pressure (mm HG)		$Kp = \bar{p}/p°$
		$\bar{p}$	$p°$	
Benzene	25	≥3.5	87	≥0.04
Benzene	25	12.0	95	0.13
Pentene-1	25	0.75	470	0.001_3
Pentene-1	50	6.25	1470	0.004_3

Note: $\Delta H = 6400$ cal (for pentene). $\bar{p}$ = Partial pressure of olefin in
equilibrium with solid complex; $p°$ = vapor pressure of pure olefin
at same temperature.

complexes is markedly higher than that of benzene-CuTFA complexes. The
difference, nearly two orders of magnitude, indicates that solvent coordination in
CuTFA/aromatic solutions is relatively weak and that the olefin-Cu(I) interactions
represented by Eq. (4) would not be adversely affected by the solvent coordination
represented by Eq. (3).

In Table 4, the dissociation constant for the pentene-1 complex of CuTFA is
compared with those reported for the solid complexes of CuCl and $AgBF_4$ [26,28].
The $AgBF_4$ is included because of its highly ionic character.

The CuCl complexes of propylene or higher olefins are unstable at 25°C, and
hence a direct comparison with CuTFA is not possible. The data in Table 5 show

Table 4 Dissociation Constants of CuTFA,
CuCl, and $AgBF_4$ Complexes with Monoolefins
and Diolefins

Salt	Olefin	$Kp = \bar{p}/p°$	
		22°C	50°C
$CuOOCCF_3$	Pentene-1	0.0013	0.0043
CuCl[a]	Ethylene	0.12	0.3
	Propylene	>1.0	>1.0
$AgBF_4$[b]	Butadiene	0.018	0.065
	Isoprene	0.2	0.9
	Ethylene	0.0003	0.006
	Butene-1	0.0043	0.011

[a]See Refs. 26 and 27.
[b]See Ref. 28.

Table 5   Solubility of CuTFA in
Nonaqueous Solvents at 23°C

Solvent	Solubility, g CuTFA/100 g solvent
Propionitrile	112
Acetonitrile	53
Xylenes	72
Benzene	37
Propylene carbonate	50
Dodecene-1	>40
Sulfolane	~13

that CuTFA is a stronger complexing agent than CuCl or $AgBF_4$. The pentene-1-CuTFA complex is about 100 times more stable than the ethylene-CuCl complex. The ethylene-CuTFA complex is expected to be over 1000 times more stable than the ethylene-CuCl complex. Similarly, extrapolation of the $AgBF_4$-olefin data suggests that the pentene-1-CuTFA complex is an order of magnitude more stable than the pentene-1-$AgBF_4$ complex.

Complexes Formed in CuTFA Solutions

Cuprous trifluoroacetate shows high solubility in a variety of solvents (see Table 5). This high solubility is due primarily to its low crystal energy. In fact, the salt sublimes at 110°C at 10^{-4} torr. The degree of association of CuTFA in benzene and in propionitrile was determined by vapor pressure osmometry (VPO) using a Macrolab model 301A apparatus. The determination consisted in comparing the "apparent molarity" obtained by VPO with the "stoichiometric molarity" of the solutions. The ratio R, stoichiometric/apparent, yields the degree of association. The results are shown in Table 6.

In benzene, R is ~3.5. Thus, in this solvent CuTFA exists as an equilibrium of tetramers with some trimers or dimers. This behavior is a characteristic of low-crystal-energy salts (such as the CuTFA) dissolved in nonpolar solvents. However, with the more polar propionitrile, the salt is essentially monomeric at moderate concentrations (0.4 M) and is partly dissociated at low concentrations (0.025 M).

By assuming that the CuTFA clusters in solution are not affected by the addition of olefin and that the only effect of adding an olefin is an increase in the apparent molarity due to free, uncomplexed olefin,* we can estimate approximate equilibrium constants.

The equilibrium constant K_{eqb} for the complexing reaction in solution:

---

*This assumption has been validated by data analysis.

Table 6 Association of CuTFA in Benzene and Propionitrile by
Vapor Pressure Osmometry

Solvent	Molarities		Ratio stoichiometric/apparent
	Stoichiometric	Apparent	
Benzene	0.3750	0.107	3.50
	0.1875	0.055	3.41
	0.1313	0.036	3.65
	0.0938	0.024	3.92
	0.0750	0.019	3.94
	0.0469	0.012	3.90
Propionitrile	0.4035	0.388	1.04
	0.2018	0.220	0.917
	0.1009	0.119	0.848
	0.0504	0.062	0.813
	0.0252	0.031	0.813

$$\underset{[Un]}{(Olefin)_{solvent}} + \underset{[M]}{(CuTFA)_{solvent}} \rightleftharpoons \underset{[M{\cdot}Un]}{(CuTFA{\cdot}olefin)_{solvent}} \tag{12}$$

is given by

$$K_{eqb} = \frac{[M{\cdot}Un]}{[Un][M]} \tag{13}$$

The equilibrium constants obtained were only used to compare the two solvents as media for olefin complexing rather than as true thermodynamic constants. The equilibrium constants in Table 7 indicate that both decene-1 and hexadecene-1 react with CuTFA in benzene to form complexes. The log K values for decene-1 are 1.9 ± 0.1. As expected, log K values for hexadecene-1 are lower, $\sim 1.3 \pm 0.1$.

In propionitrile, decene-1 forms complexes, whereas hexadecene-1 barely interacts with CuTFA at the low salt concentration used to keep within the range valid for VPO measurements. At higher salt concentrations, hexadecene-1 would complex also, subject to solubility limitations. The effect of salt concentration is particularly important with strongly coordinating solvents capable of competing with the double bond for Cu(I) coordinating sites. Calorimetric determinations have addressed this question.

Calorimetry of Cuprous Complexes in Solutions: Solvent and Anion Effects

An investigation of the solution equilibria involved in Cu(I)-hexane interactions was made using thermometric titration. The technique has been successfully used

Table 7 Approximate CuTFA-Olefin Complexation Equilibria by Vapor Pressure Osmometry

Solvent	Olefin	Stoichiometric molarity		Apparent molarity	Expected[a] CuTFA molarity	Free olefin	K_{eqb}
		CuTFA	Olefin				
Benzene	Decene-1	0.399	0.1055	0.127	0.122	0.005	70
		0.200	0.0528	0.064	0.060	0.004	81
	Hexadecene-1	0.100	0.020	0.036	0.028	0.008	17
		0.200	0.120	0.100	0.060	0.040	17
		0.100	0.060	0.052	0.028	0.024	24
		0.100	0.100	0.078	0.028	0.050	20
Propionitrile	Decene-1	0.4035	0.400	0.446	0.388	0.058	95
		0.2018	0.200	0.255	0.220	0.035	13
	Hexadecene-1	0.2012	0.040	0.262	0.220	0.042	1.0
		0.2047	0.1601	0.362	0.220	0.142	0.7

[a]Based on solvent-CuTFA VPO data.

in the determination of pK values for proton ionization in aqueous solution [30], and for the determination of log K values for metal-ligand interactions in aqueous and nonaqueous solvents [31,32].

In this work, thermometric titration data were accurate to within 2–4%. This margin of error could cause a significant uncertainty of the log K values for reactions with small $\Delta H°$ values. However, significant trends can still be recognized, particularly when the Cu(I)-olefin interactions is strong.

Results for the interactions of various cuprous salts dissolved in toluene, propionitrile, and propylamine with hexene-1 are shown in Table 8. These results show that the chloride does not complex hexene-1 in propionitrile. In toluene and in propionitrile, CuTFA complexes only one hexene-1 molecule, whereas DTBS and the chloroalanate complex two hexenes. In toluene, the more ionic salts CuTFA and $CuAlCl_4$ show a higher energy of interaction in the first olefin-complexing step (about 10 kcal/mol) than the more covalent CuDTBS (6 kcal/mol). This result suggests that the cuprous ion in the more ionic cuprous salts is more available for interaction with the olefin double bond than the cuprous ion in the more covalent cuprous salt.

Table 8 Log K, $\Delta H°$, and $\Delta S°$ Values for the Reaction CuA + $nC_6^= = CuA(C_6^=)_n$ at 30°C

A	n	Log K	$\Delta H°$ (kcal/mol)	$\Delta S°$ (e.u.)
Toluene Solvent				
TFA⁻	1	2.6	−10.6	−23
DTBS⁻	1	1.6	−6.0	−13
	2	3.1	−13.3	−30
AlCl₄⁻	1	1.6	−9.8	−25
	2	2.6	−13.7	−34
Propionitrile Solvent				
Cl⁻			0	
TFA⁻	1	0.6	−1.2	−1
EtSO₃⁻	1	1.6	−0.5	6
ΦSO₃⁻	1	1.0	−0.3	4
AlCl₄⁻	1	0.6	−1.6	−3
	2	1.2	−0.4	4
DTBS⁻	1	0.8	−1.3	0
	2	0.5	−5.0	−15
Salic⁻	1	1.2	−0.7	3
	2	0.7	−2.7	−6
n-Propylamine Solvent				
TFA⁻			−0.2	
DTBS⁻	1	0.4	−0.8	−1

The propionitrile-cuprous salt data allow us to order the various anions in the order of increasing ionicity (or strength of interaction) as follows:

Chloride < acetate < ethylsulfonate ~ phenylsulfonate <
di-*tert*-butyl salicylate ~ salicylate < trifluoroacetate < chloroalanate

Finally, competition between olefin and solvent for cuprous coordination sites increases in the order:

Toluene < propionitrile < propylamine

The interaction of the amine solvent with the Cu(I) salts is too strong for practical use of an amine solvent system for monoolefin recovery. Interaction of the amine with the Lewis acid, $AlCl_3$, is so large that the following reaction occurs:

$$CuAlCl_4 + PrNH_2 = CuCl + PrNH_2AlCl_3 \qquad (14)$$

The thermometric titrations have been carried out using low concentration of cuprous ion in the solution, about 1 wt % in Cu(I). The fact that the conclusions derived at these low concentrations support our general thesis about the behavior of ionic cuprous-olefin complexes is gratifying.

PROCESS IMPLICATIONS

Separation of Low Molecular Weight Olefins from Paraffins

The high affinity of ionic cuprous complexes for light olefins makes the separation of light olefins from paraffins technically feasible. In fact, with ionic cuprous salts dissolved in nonaqueous solvents, it is possible to separate ethylene and propylene from ethane and propane in one step, a separation, which is not possible by conventional distillation.

In conventional distillation, the ethylene is more volatile than ethane, and propylene more volatile than propane. The relative volatilities for these pairs at 0°C are low: $\alpha = 1.7$ for the ethylene/ethane pair and 1.2 for the propylene/propane pair. In extractive distillation using ionic cuprous solutions, the olefins become the less volatile components. Complexing with Cu(I) lowers the activity coefficient of the olefins, and salting out by the polar salt raises the activity coefficient of the paraffins. Both effects become more important with increasing salt concentration, and hence both selectivity and capacity increase with increasing salt content.

The solubility of CuTFA in both aromatic and nitrilic solvents makes it possible to compare the effect of solvent coordination on the complexing ability of the cuprous ion. The propylene/propane capacity ratios for CuTFA/xylene and CuTFA/Etcn are shown in Figure 1. At olefin/Cu(I) ratios less than 0.3, the propylene/propane capacity ratios have been found to be almost identical to the selectivities determined in experiments using olefin/paraffin mixtures as feeds. At higher loadings of 1 mol of olefin per Cu(I), the experimental selectivities obtained

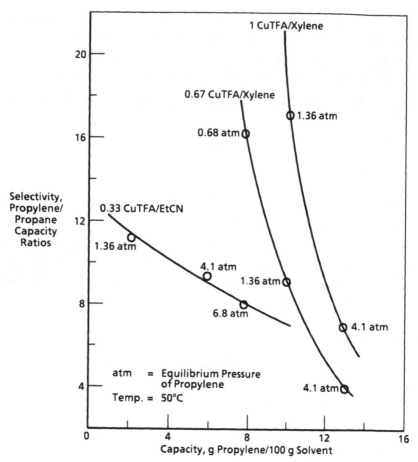

Figure 1   Capacity-selectivity of CuTFA/xylene, CuTFA/EtCN, and CuTBS/EtCN systems.

with mixed feeds are 30–40% lower than the capacity ratios. In general, however, trends obtained with capacity ratios are followed by selectivities determined with olefin/paraffin mixed feeds.

Figure 1 shows that the capacity ratios at a given solvent loading are higher for the 1 CuTFA/xylene system than for the 0.67 CuTFA/xylene system. This figure also shows that at high equilibrium pressures of propylene, i.e., 100 psia, the selectivity afforded by the CuTFA/EtCN system is higher than the selectivity of the CuTFA/xylene systems. In fact at equilibrium pressures in excess of 1.5 atm, the selectivity of CuTFA/xylene systems declines rapidly. Chemical interactions predominate at low loadings, but physical solubility effects become important at high loadings. Thus, the nonpolar, low-dielectric-constant aromatic solvent (which allows clustering of CuTFA species) results in low selectivities at high loadings. The propylene and propane isotherms from 1 CuTFA/xylene solution are shown in Figure 2.

These trends suggest that aromatic solvents can be useful as carriers of ionic cuprous salts in applications where the olefin concentration in the feed is at relatively low partial pressure, under 3.5 atm, but propionitrile would be the preferred solvent for olefin recovery and purification from streams in which olefin occurs at high concentrations and partial pressures. Thus, the aromatic-based systems should be considered for the recovery of olefins from waste gas streams, whereas the more polar nitrilic based-systems are more appropriate for the recovery of olefins from olefin-rich pyrolysis gas streams.

The recovery of the olefin from the cuprous salt complex by heating and depressurization is easier from the propionitrile system. This is shown in Table 9, which compares the equilibrium constants of propylene at 75°C for CuTFA/xylene and CuTFA/EtCN systems.

The high cost of CuTFA salt reduces its attractiveness for large-scale separations. Alternative salts, such as $CuNO_3$ and Cu_2SO_4, offer more economical options. The volumetric capacities of $CuNO_3$, CuTFA, and Cu_2SO_4 solutions in EtCN for ethylene, propylene ethane, and propane are summarized in Table 10.

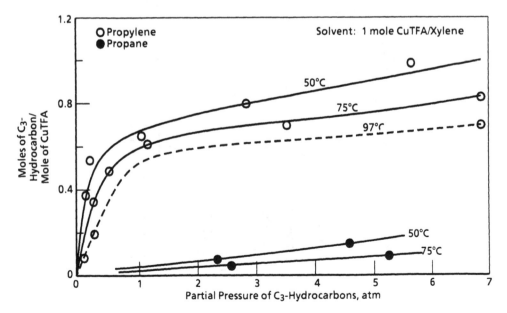

Figure 2 Propylene and propane isotherms from 1CuTFA/xylene solution.

Table 9 Volumetric Capacities[a] of CuNO$_3$ and CuTFA Solutions for Ethylene Propylene, Ethane, and Propane

Solvent	wt % Cu(I)	g Cu(I), l	Temp. (°C)	1 atm		5.5 atm		8.2 atm			
				C$_2^=$	C$_3^=$	C$_2^=$	C$_3^=$	C$_2^=$	C$_3^=$	C$_2$	C$_3$
CuNO$_3$/EtCN	18.2	210	25	18	9	41	35	47	43	6.6	7.7
>>	22.6	315	25	30	29	57	61	68	79	~1	~1
CuTFA/EtCN	19.0	230	25	25	25	63	63	74	75	—	10
Cu$_2$SO$_4$/EtCN	17.5	220	50	8	—	32	—	36	—	—	—
CuNO$_3$/EtCN	18.2	210	50	15	5	34	22	40	29	5.8	5.5
>>	22.6	315	50	23	16	47	47	54	58	~1	~1
CuTFA/EtCN	19.0	230	50	15	14	49	44	68	54	—	6
CuNO$_3$/EtCN	18.2	210	75	13	3	25	11	31	14	4.2	4.5
>>	22.6	315	75	16	9	34	29	39	37	<1	1
CuTFA/EtCN	19.0	230	75	9	7	21.4	16	44	27	~1	~1

[a]Volumetric capacities are given in liters of gas measured at 25°C, 1 atm per liter of solution. The uncertainty is ±4% of the reported value, due to nonideality corrections, interpolation, experimental precision, and approximations in estimating partial molal volumes of complexed olefins.

Table 10 Equilibrium Constants of Propylene at
75°C from Various CuTFA Solutions

Moles of solvent / Moles of CuTFA	P̄ Propylene (psia)	K_{eqb}[a]
1Xylene	5	3.4
	20	2.3
1.5Xylene	5	4.5
	20	3.1
3EtCN	5	50.0
	20	19.0

[a]K equilibrium = $\dfrac{\text{wt\% propylene in vapor}}{\text{wt\% propylene in liquid, CuTFA-free}}$

This table shows that with such systems a one-step separation of ethylene and propylene from ethane and propane is feasible. The capacity ratios, and hence the selectivities, increase with salt concentrations. Thus, $CuNO_3$/EtCN containing 18.2 wt % Cu(I) exhibits selectivities for ethylene/propane of about 5. At 22.6 wt % Cu(I), the selectivities rise to about 50.

A process which has been proposed for the recovery of ethylene from waste gases is the ESEP process. This process, demonstrated by Tenneco [25], uses $CuAlCl_4$ in heavy aromatic solvents. An important technological achievement in this process is the inhibition of Friedel–Crafts-type side reactions which can be catalyzed by the Lewis acid component $AlCl_3$ in the chloroalanate ion. Such reactions would lead to the addition of ethylene to the aromatic solvent.

An attractive application of the ionic cuprous salt/nonaqueous solvent systems is in propylene/propane separation. In the conventional distillation, this separation requires large reflux ratios and a large number of theoretical plates because the propylene/propane pair has a very low relative volatility. The ionic cuprous salts discussed in this section offer the possibility of an extractive distillation process, which would significantly reduce the energy requirements of the separation. A simplified flow scheme of this concept is shown in Figure 3. Another potential application is the simultaneous removal of ethylene and propylene from olefin plant cracked products containing hydrogen and methane along with the C_2-C_3 fraction. The C_2-C_3 fraction contains some acetylenes in addition to the olefins and paraffins. The comparative economics of the conventional and the cuprous complex cases depend on the separation and purification steps used to handle the hydrogen, methane, acetylenes, carbon monoxide, and heavier hydrocarbons. Thus, in the cuprous process it is not necessary to remove hydrogen by refrigeration before the olefin recovery. This minimizes the requirements for deep refrigeration needed to affect the separation of hydrogen gas from the total stream.

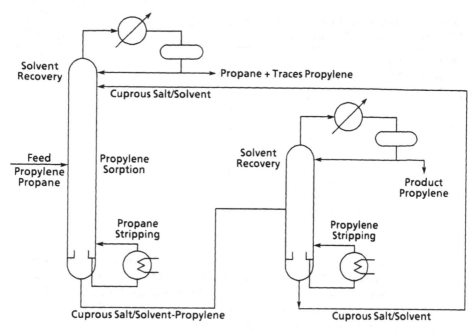

Figure 3 A generalized extractive distillation scheme for propylene/propane separations.

The regeneration temperature in an ionic cuprous salt/solvent system is an important parameter. The rates of thermal decomposition of CuTFA and $CuNO_3$ are discussed in Appendix III. To avoid thermal decomposition, reboiler temperatures in CuTFA systems should be 100°C or less. For $CuNO_3$/EtCN systems, thermal stability considerations dictate reboiler temperatures of 95°C or less. Finally, in designing separations based on cuprous salt systems, we must also take into account the chemical reactivity of the cuprous ion toward oxygen, water, and sulfur compounds. Thus, feed pretreatment is necessary to minimize chemical decomposition of the cuprous system.

Separations of High Molecular Weight Olefins From Paraffins

The system CuTFA/EtCN is sufficiently polar to be essentially immiscible with high molecular weight paraffins. The polarity of the solution increases with salt concentration and along with it the selectivity of the solution for olefins increases. This characteristic of the CuTFA/EtCN system permits its use in a solvent extraction process for high molecular weight olefin recovery.

The selectivities of the CuTFA/EtCN system for decene/decane separations are compared with the selectivities of various other cuprous salt/nonaqueous

Table 11 Selectivities of Various Cuprous Salt/
Nonaqueous Solvent Systems for Decene-1/Decane
Separations by Solvent Extraction at 25°C

Solvent system	Olefin loading of polar phase, wt % Salt-Free Basis	β value C_{10}/C_{10}
1CuTFA/2EtCN	11.0	34.0
1CuTFA/3EtCN	9.5	19.0
1CuTFA/3BuCN[a]	11.4	9.7
1CuTFA/3ProNH$_2$[a]	4.9	2.1
1CuTBS/5.7EtCN	8.0	2.5
1CuAlCl$_4$/5EtCN	3.1	2.6
1CuΦSO$_3$/9ProNH$_2$[a]	6.7	1.1
1CuSalic ProNH$_2$[a]	4.5	2.2

[a]ProNH$_2$ = propylamine; BuCN = butyronitrile.

solvent systems in Table 11. The olefin capacity and selectivity of the CuTFA/
EtCN system is clearly higher than it is for the other salt/solvent systems shown.
The reduction in selectivity in going from EtCN solvent to BuCN solvent is due to
the lower polarity of the latter solvent. The use of propylamine is detrimental due
to competitive coordination effects. Addition of the polar cosolvent sulfolane to
CuTFA and CuDTBS solutions in EtCN increases selectivity at the expense of
capacity (Table 12).

Selectivities (β) of the CuTFA/3EtCN system for α-dodecne/dodecane and
for α-tetradecene/tetradecane pairs at 25°C range from 4 to 27 depending on
solvent loading (see Figures 4 and 5). Selectivities increase with increasing salt
concentrations but remain satisfactory even at salt/solvent ratios at 1:5 and at

Table 12 Effect of Addition of Sulfolane on Selectivity of CuDTBS
and CuTFA Systems at 25°C

Solvent system	Sulfolane, wt %[a]	Olefin loading of polar phase, wt % Salt-Free Basis	β value C_{12}/C_{12}
1 CuDTBS/5.6EtCN	—	7.8	2.5
1 CuDTBS/5.6EtCN	10	7.2	3.7
1 CuTFA/5.0EtCN	—	3.5	12.7
1 CuTFA/5.0EtCN	10	3.3	16.0

[a]Sulfolane wt % basis polar phase.

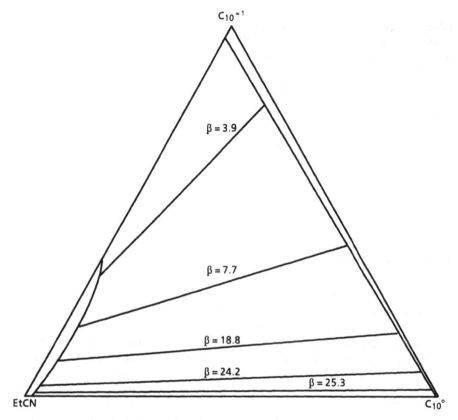

Figure 4 Selectivities (β) of 1CuTFA/3EtCN for decene-1/decane at 25°C (composition in weight fractions, CuTFA-free basis).

temperatures up to 50°C. Thus, a one-step separation of C_{10}-C_{14} terminal olefins from the corresponding paraffins by solvent extraction using CuTFA/EtCN systems is feasible. Back extraction of the complexed olefins can be affected by contacting with a lighter molecular weight paraffin, such as octane (Figure 6) or pentane (Figure 7).

The performance of CuTFA/EtCN in separations of linear internal tetradecenes is disappointing. The capacity of CuTFA/2EtCN solvent for mixed tetradecenes is only 5 wt % basis total solvent compared to over 25 wt % for tetradecene-1, and the β value at 50°C and 1 wt % loading is ~ 7 compared to over 20 for tetradecene-1/tetradecane.

With regard to plasticizer olefins (C_6-C_9) the β values of a CuTFA/2EtCN solvent for heptene-1/heptane at 50°C range from 3 at wt % olefin loading to 50 at 2

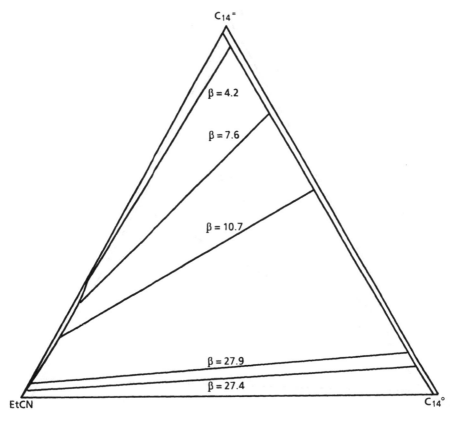

Figure 5 Selectivities (β) of 1CuTFA/3EtCN for tetradecene-1/tetradecane at 25°C (composition in weight fractions, CuTFA-free basis).

wt % (Figure 8). Again the selectivities for internal olefins are significantly lower (Figure 9).

The data presented in this section suggest that although terminal olefins can be efficiently recovered from paraffins both in the plasticizer (C_6-C_9) and in the detergent range (C_{10}-C_{15}), internal olefins would be only partially recovered. Processes for the separations of terminal from internal olefins are, however, feasible and could be of commercial interest. For such processes, higher EtCN/CuTFA ratios would be used to increase the competition between the internal olefins and the solvent for the Cu(I) coordination site.

An ambient temperature isothermal process for the recovery of terminal detergent range olefins from paraffins is shown in Figure 10. The olefins are extracted at 25° to 40°C using a rotary disk contractor to contact the cuprous

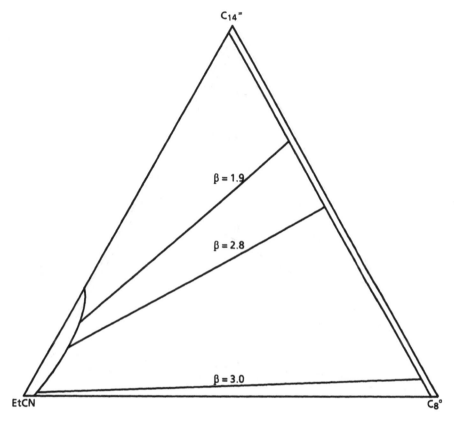

Figure 6   Selectivities (β) of 1 CuTFA/3EtCN for tetradecene-1/Octane at 25°C (composition in weight fractions, CuTFA-free basis).

solvent with the feed. The stripping of the olefins is carried out at ambient or higher temperatures by contacting with a lighter paraffin. Two distillation steps— one for the recovery of EtCN from the paraffinic raffinate, the other for the recovery of the paraffinic countersolvent from the olefin product—are also needed. On the basis of Ponchot–Savarit calculations, the recovery of 90 wt % dodecene/10 wt % dodecane product from a 20 wt % dodecene-1/80 wt % dodecane feed can be carried out at solvent/feed ratios of 2.2 at a reflux to the enriching section of about 2 (basis product). The overall solvent ratio at this reflux is 6. Four stages in the recovery section and two stages in the enriching section are needed. For the olefin-stripping section, at a solvent/feed ratio of 0.8 (where

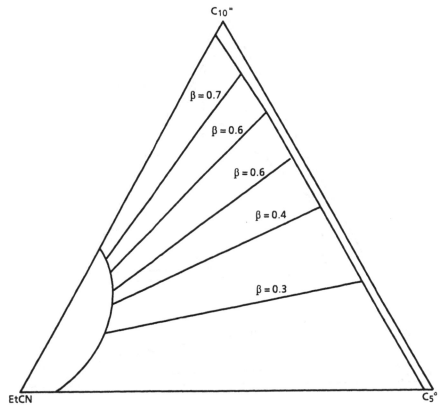

Figure 7 Selectivities (β) of 1CuTFA/3EtCN for decene-1/pentane at 25°C (composition in weight fractions, CuTFA-free basis).

solvent is octane and feed is the weight of the fat solvent, EtCN/CuTFA plus olefin), five stages would be required (see Table 13). Stripping requirements can be decreased by increasing the stripping temperature to 70 or 80°C.

The propionitrile overhead from distillation column D is returned to the upper sections of the extractors RDC-A and RDC-B to wash the CuTFA from hydrocarbon streams leaving these units.

The CuTFA dissolved in the raffinate phase can also be removed by sorption on an active carbon bed. Total removal of 0.5 wt % CuTFA from 11% dodecene-1 feed was realized in an active carbon bed. The sorbed salt can be recovered from the bed by an EtCN wash.

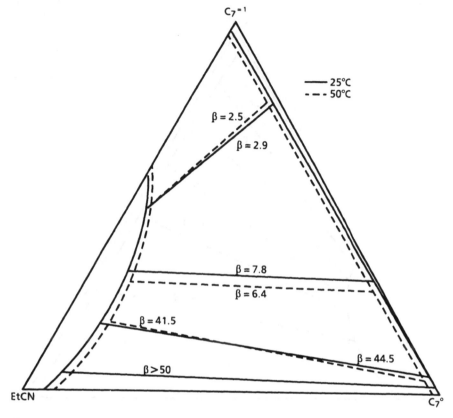

Figure 8 Selectivities (β) of 1CuTFA/2EtCN for heptene-1/heptane at 25 and 50°C (composition in weight fractions, CuTFA-free basis).

Styrene/Ethylbenzene Separation Based on Dual-Solvent Cuprous Salt/Propionitrile/Paraffin Systems

An olefin/paraffin separation of industrial importance is the purification of styrene (Sty) from ethylbenzene (EB)/styrene feeds. Present commercial practice of this separation [33] involves high-temperature (100–110°C) vacuum distillations. The α value for the ethylbenzene/styrene pair at these conditions is about 1.3. Thus, stage and reflux requirements for the recovery of 98% of the styrene at 99.6% purity are very high.

     Two cuprous salt systems have been considered for this separation: CuTFA/EtCN and $CuNO_3$/ETCN [18]. Both Sty and EB are completely miscible in CuTFA/EtCN solutions containing 2 or 3 mol of EtCN per mol of cuprous salt. In $CuNO_3$/EtCN systems, the Sty is again completely miscible, but EB is only

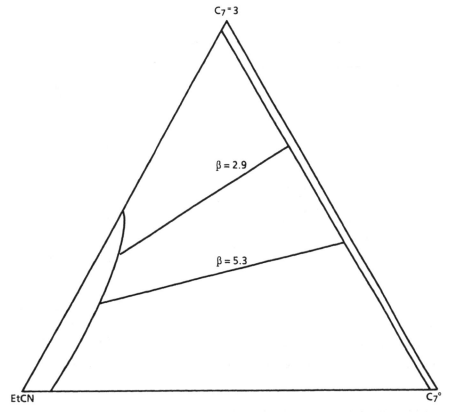

Figure 9 Selectivities (β) of 1CuTFA/2EtCN for heptene-3/heptane at 50°C (composition in weight fractions, CuTFA-free basis).

soluble up to 20 or 30 wt %. In either case, a countersolvent which is immiscible with the cuprous salt solution is required for an effective solvent extraction separation.

The selectivities for the Sty/EB separation exhibited by the two solvent systems are shown in Table 14. In order to keep the viscosity of the system low, the CuTFA system would operate at a minimum temperature of 50°C; thus the data for CuTFA are given at 50°C. The minimum temperature in the $CuNO_3$ system is 40°C, limited by the temperature of cooling water. n-Tetradecane is used as the countersolvent because lighter paraffins yield considerably lower selectivities with the CuTFA systems. The following conclusions can be drawn from data in Table 14:

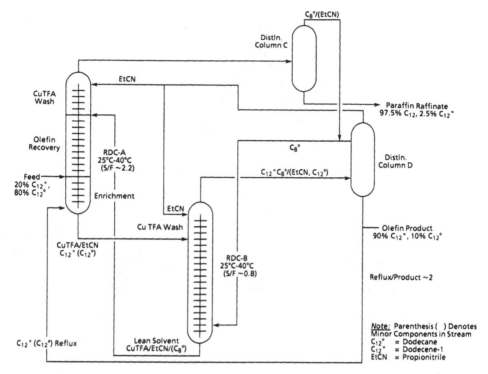

Figure 10   Isothermal process for the recovery of detergent range olefins from paraffins.

Table 13   Estimated[a] Staging and Solvent Requirements for
Dodecene-1/Dodecane Separations Using CuTFA/3EtCN Solvents

Conditions
Feed: 20% Dedecene-1 in dodecane
Purity: 90% Dodecene-1 in dodecane
Recovery: 90% of dodecene-1

Olefin Extraction (25°C)
S/F weight ratio ~2.2
Reflux to enriching section ~2 (basis product)
Minimum reflux ~1
Stages: Recovery section 4
              Enriching section 2
Weight of dodecene recovered per 100 w of CuTFA circulated: 6.5

Olefin Stripping (25°C)
S/F weight ratio ~0.8
Stages ~5

---

[a]Based on Ponchot–Savarit graphical constructions.

Table 14 Comparison of Selectivities (β) of CuNO$_3$/EtCN/n-Tetradecane and CuTFA/EtCN/n-Tetradecane in Styrene/Ethylbenzene Separations

Temp. (°C) of CuNO$_3$/EtCN system[a]	Sty/Cu(I)	wt % nC$_{14}$ in nonpolar phase	Selectivity, β[b]	
			CuNO$_3$/EtCN 22 wt % Cu[c]	CuTFA/EtCN 19 wt % Cu[c]
25	0.44	80.2	6.8	—
50	0.44	72.2	4.3	2.7
25	0.36	81.6	7.2	—
50	0.36	87.0	7.6	5.3
40	0.61	62.0	4.0	<2.1
40	0.52	67.0	4.2	<2.2
40	0.20	91.0	11.5	7.5
40	0.10	93.6	>50.0	12.0

[a]All CuTFA/EtCN selectivities refer to 50°C data.

[b]Selectivity $\beta = \dfrac{\dfrac{\text{wt\% Sty in polar phase}}{\text{wt\% Sty in nonpolar phase}}}{\dfrac{\text{wt\% EB in polar phase}}{\text{wt\% EB in nonpolar phase}}}$

[c]Saturated solutions at room temperature.

1. At comparable conditions of temperature, loading, and countersolvent concentrations, the selectivities of the CuNO$_3$ solvent are 40–60% higher than those of the CuTFA solvent. Due to the higher Cu(I) content and density of the CuNO$_3$ system, the solvent flow rate at equal Sty/Cu(I) ratios is 25% lower for the CuNO$_3$ system.
2. The selectivities at low Sty/Cu(I) ratios (~0.1) are considerably higher in the CuNO$_3$/EtCN system. Thus, high purities can be reached with fewer stages using CuNO$_3$/EtCN.

The above discussion shows that the CuNO$_3$/EtCN system is superior to CuTFA/EtCN under the most favorable conditions for the latter system, i.e., with n tetradecane countersolvent. The effect of molecular weight of countersolvent on the selectivity and countersolvent requirements of CuNO$_3$/EtCN/paraffin systems is also important. In Table 15, selectivities of CuNO$_3$/EtCN with tetradecane and with heptane are compared with selectivities obtained with CuTFA/EtCN/n-paraffin systems. This table shows that with CuNO$_3$/EtCN, n-heptane is even more effective as a countersolvent than n-tetradecane. This trend is opposite that observed with CuTFA/EtCN systems, and is attributed to the high polarity of CuNO$_3$ relative to CuTFA. The high polarity of CuNO$_3$/EtCN results in heptane solubilities lower than 0.5 wt %, whereas the solubility of heptane in CuTFA/EtCN is in the order of 5 wt %.

Table 16 compares the countersolvent concentration (wt %) in the non-polar

Table 15 Effect of Molecular Weight of Paraffin Countersolvent on the Selectivity of CuNO₃/EtCN and CuTFA/EtCN for Styrene/Ethylbenzene Separations

Solvent	Countersolvent	Sty/Cu(I)	wt % Paraffin in nonpolar phase	Selectivity, β[a]
CuNO₃/EtCN	n-Tetradecane	0.20	91	11.5
22wt % Cu(I)	n-Tetradecane	0.50	67	4.2
	n-Heptane	0.16	88	14.4
	n-Heptane	0.53	59	5.7
	n-Heptane	0.50	67	Over 7 (interpolated)
CuTFA/EtCN	n-Tetradecane	0.20	91	7.5
19wt % Cu(I)	n-Tetradecane	0.50	67	<2.0
	n-Decane	0.20	85	5.7
	n-Decane	0.50	85	2.0

[a]Selectivities of CuNO₃/EtCN are at 40°C; those of CuTFA/ETCN at 50°C. These are optimal temperatures for both systems.

phase, at which a selectivity of 4 is achieved with the various systems, at a Sty/Cu(I) molar ratio of 0.5–0.6. From these concentrations, the corresponding mass flow rates are calculated and shown in the last column of the same table. This table shows that a 10-fold reduction in countersolvent flow rate is possible in going from the best CuTFA/EtCN case with n-tetradecane to CuNO₃/EtCN with n-heptane.

A preliminary process flow scheme for the separation of 230 M metric tons/year of Sty from EB is shown in Figure 11. The feed is the crude product from a dehydrogenation plant and consists of 35 wt % EB and 65 wt % Sty. The conventional process requires two large vacuum distillation columns (35 mm Hg) in parallel, each with 70 trays, 7-m top section diameter and 5.5-m bottom section diameter, followed by a second vacuum distillation (32 mm Hg) in a 6.4-m-diameter column with 20 trays.

Table 16 Concentrations of Countersolvent Required to Yield β±4 with Cu(I)/EtCN/Paraffin Systems Styrene/Cu(I) Molar Ratios 0.5–0.6

System[a]	wt % Paraffin counter-solvent in nonpolar phase	Weight ratio counter-solvent feed components
CuNO₃/EtN/n-Heptane	45	<1
CuNO₃/EtCN/n-Tetradecane	67	2
CuTFA EtCN/n-Tetradecane	90	10
CuTFA/EtCN/n-Decane	95	20

[a]CuNO₃ data at 40°C, CuTFA data at 50°C. These are optimal temperatures for both systems.

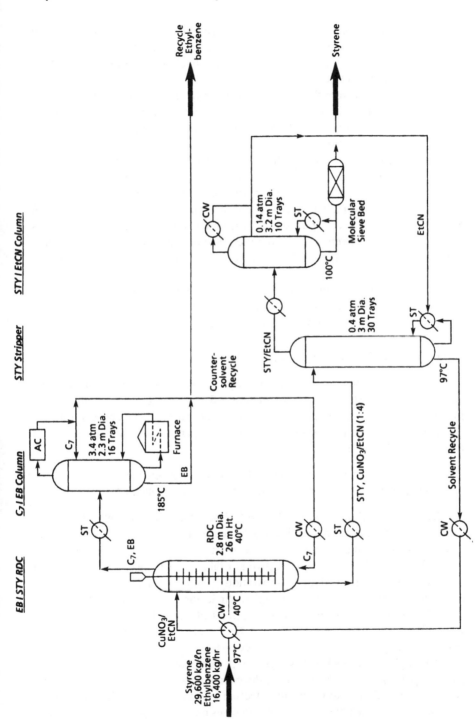

Figure 11 Styrene recovery from EB/Sty by dual-solvent liquid–liquid extraction (230 million kg/yr Sty).

In the CuNO$_3$/EtCN process 230 M metric tons/year of Sty is separated from EB by liquid–liquid extraction in a rotary disk contactor (RDC). The extraction process is carried out at 40°C at a pressure slightly above atmospheric. An RDC 2.8 m in diameter and 26 m high is needed.

EB is separated from the C$_7$ countersolvent by distillation at 185°C and 3.4 atm. The EB bottom is recycled to the dehydrogenation reactor. The counter-solvent overhead is recycled to the RDC. The C$_7$/EB distillation requires 16 trays and a reflux-to-feed ratio of 0.8.

The solvent stripper removes Sty from the CuNO$_3$/EtCN solvent. In order to minimize polymerization of Sty and decomposition of the solvent, the maximum temperature of the stripper is limited to 95–100°C. The bottom product, CuNO$_3$/EtCN, is cooled and recirculated to the RDC. The overhead product, 70 wt % Sty/30 wt % EtCN, is separated by vacuum distillation. EtCN overhead is recirculated to the solvent stripper reboiler. The bottoms from this distillation is the Sty product. Alternatively, Sty can be extracted from the cuprous solvent in a heavier paraffinic countersolvent.

A process using cuprous dodecylbenzenesulfonate dissolved in aromatic solvents for the recovery of styrene from a mixture of closely boiling hydro-carbons (not ethylbenzene) has also been proposed [34].

Separations of Diolefins by Cuprous Complexing

The use of nonaqueous solutions of ionic cuprous salts for the recovery of isoprene has been patented [15,19,20]. The relative affinities of the cuprous salt in such systems for the various unsaturates follow the general trend one expects from the order of Cu(I) double-bond stabilities. Thus for C$_5$ unsaturates, the bond strength decreases in the order

isoprene > pentene-1 > 3-methyl butene-1 > *cis*-pentene-2 > *trans*-pentene-2 > > methyl butene-1 > 2 methyl butene-2

Relative volatilities from CuTFA/EtCN at infinite dilution and 30°C were calculated from the emergence times obtained from gas–liquid chromatography columns in which the packing was coated with CuTFA/5EtCN solutions. The results for various isoprene pairs are given in Table 17. A more economical system which has also been studied for the separation of butadiene and isoprene is CuNO$_3$/EtCN. Selectivities for butadiene at high olefin loadings are given in Table 18. The critical separation in this process is between butadiene and butene-1, and relative volatilities for this pair are 1.7–1.9 at 65°C. With high volumetric capacities this system is competitive with existing processes.

Although the CuNO$_3$/EtCN system is competitive with processes using physical polar solvents (DMF, NMP), the economic advantage of the cuprous salt system is not sufficient to justify its selection over the better established physical solvent systems. The physical solvent processes have been optimized over the last

Table 17 Relative Volatilities at
Infinite Dilution of Amylene/Isoprene
Pairs from CuTFA/5EtCN at 30°C

Solute	Relative volatility
Isoprene	1.0
Pentene-1	3.5
3-Methyl butene-1	4.0
2-Methyl butene-1	7.0
2-Methyl butene-2	15.0

Table 18 Relative Volatilities of Butene-1/
Butadiene from CuNO$_3$/EtCN Solutions at 65°C

Solution concentration (wt %)			Double bond/Cu(I)	α
Cu(I)	Bu-1	BD		
15	15.4	3.63	2.1	1.73
15	2.3	17.6	3.3	1.88

Note: Equilibrium pressure of olefins: 4 atm.

20 years by implementing energy-saving modifications, thus reducing the incentive to develop the untried cuprous salt process.

In considering separations of diolefins, we should keep in mind that some diolefins form solid complexes in some of the systems we have been discussing. Thus, in CuTFA/aromatics and in Cu$_2$SO$_4$/EtCN, symmetric diolefins tend to make insoluble complexes at high loadings. Such complexes have not been observed in CuTFA/EtCN or CuNO$_3$/EtCN systems.

Recovery of Carbon Monoxide by Cuprous Salt Solutions

Even though carbon monoxide is not an unsaturated hydrocarbon, a discussion of the process capabilities of cuprous salts would not be complete without reference to the use of cuprous systems in the recovery of CO.

The cuprous complexes of CO are very stable. Thus, aqueous cuprammonium acetate has been used industrially to complex and recover carbon monoxide. However, with the increased understanding of the complexing potentialities of ionic cuprous salt/nonaqueous solvent systems, the development of COSORB process [28] evolved naturally in the early 1970s. This process, like the ESEP process already discussed, uses the ionic salt CuAlCl$_4$ in aromatic solvents, to complex and recover CO. The equilibrium and kinetics of this system have been discussed in detail [35].

The cuprous salt systems discussed in this paper also exhibit significant capacities for CO complexing. An interesting system not mentioned so far is CuCl/EtCN. It was not mentioned because CuCl is viewed as a covalent salt, at least in the solid state. In the solid state the CuCl is polymeric, with a network of chloride ions bridging the Cu(I) and rendering it relatively inaccessible to complexing.

The solubility of CuCl in the coordinating solvent EtCN is extremely high, however, up to 60 wt % at 25°C. Such a solution contains 40 wt % cuprous ion and can complex carbon monoxide [36], ethylene, and propylene. The ability of the CuCl/EtCN system to complex propylene can be understood in terms of what (we believe) happens in the EtCN solvent. Vapor pressure measurements of EtCN solutions of CuCl suggest that in this solvent, CuCl occurs as a dimer, $(CuCl)_2$, a much freer form of CuCl than we encounter in the polymeric, $(CuCl)_x$, form, where x is very large. Thus, Cu(I) from CuCl/EtCN solutions can interact with propylene more effectively than it can from the solid state. Even so, comparison of the CO/Cu(I) ratios obtained in CuCl/EtCN and $CuNO_3$/EtCN systems under similar conditions shows that the more ionic $CuNO_3$ is the more effective CO complexor. Thus at 25°C and 40 psi CO pressure, a solution of CuCl/EtCN containing 25 wt % Cu(I) absorbs 0.3 mol of CO per mol of cuprous ion, whereas a solution of $CuNO_3$/EtCN containing 22 wt % Cu(I) absorbs 0.7 mol CO per mol of Cu(I).

SUMMARY AND CONCLUSIONS

Ionic cuprous salts of strong organic and inorganic acids can interact with the double bond of an olefin far more strongly than the more covalently bound cuprous salts. Thus, whereas dry CuCl cannot complex propylene at ambient temperatures, dry CuTFA can complex olefins ranging from ethylene up to hexadecene-1. The complexing ability of ionic cuprous is maintained in the solution phase, provided that the solvent does not coordinate the Cu(I) ion so strongly as to interfere with the formation of olefin complexes. The fact that dissolved cuprous ion retains its affinity for olefin is particularly significant when considering separation applications, because it means that we can use cuprous salts which may not be stable in the dry form but which meet the criteria of ionicity and low cost. Two examples of such salts are $CuNO_3$ and Cu_2SO_4.

Ionic cuprous salts dissolved in nonaqueous systems can be used in many olefin separation schemes. In each application the salt/solvent combination can be tailored to fit the needs of the specific separation. Several examples of process concepts have been given. The physical properties of the solutions are favorable for process application. Their high density (1.0–1.3 g/cm³) combines with high affinity for olefins to yield excellent volumetric capacities. Their viscosities are within acceptable ranges. Their thermal stabilities are satisfactory as long as regeneration temperatures above 100°C are avoided. The corrosion rates of the

$CuNO_3/EtCN$ are extremely low, and the corrosions of CuTFA and Cu_2SO_4 systems can be controlled by appropriate means [37].

The primary impetus for the study and development of such separating media has been energy savings. With cuprous complexing systems, separation energy can be saved by minimizing refrigeration costs and by reducing reflux requirements through the use of high selectivities. Capital costs are reduced by using small contacting units made possible by the high volumetric capacities. These advantages can be achieved at the expense of a more complex plant. Moreover, the chemical stability of the solvent requires careful control of feed purity, removal of water and sulfides, and exclusion of air. These reasons, along with the economic reality of low-energy costs in the USA, have inhibited acceptance and implementation of cuprous salt/solvent systems in separations.

To facilitate the implementation of cuprous salt/nonaqueous solvent technology, it is desirable to identify applications for which there are no efficient alternative processes, and which entail small to medium-sized processing units. Such opportunities may develop in retrofitting existing plants as a means of improving product purity or to increase yields. Energy-intensive separations would be a good area of opportunity in a scenario of rising energy costs. Once the area of opportunity and best process fit have been identified, however, it is important to carry the development effort far enough along to permit optimization of the new process and to minimize the margins of uncertainty in the economic evaluations. Uncertainties in any new technology weigh the decision in favor of the established technology.

ACKNOWLEDGMENTS

The author thanks many coworkers on this project, including E. R. Bell, A. K. Dunlop, D. J. Eatough, and R. S. Slott. Thanks also to Shell Development Company for the assistance extended in the preparation of this chapter.

APPENDIX I: DETERMINATIONS OF Cu(I) AND Cu(II)

The determination of the cuprous and cupric content of solution samples was carried as follows: First, the total copper was determined by the spectrophotometric method using "neocuprine" (2,3-dimethyl-1,10-phenanthroline). Then, the cupric content was obtained by reacting a different portion of the same solution with an acidified iodide solution to form iodine. The iodine was titrated with standard thiosulfate solution to a potentiometric end point. The reaction was carried out under nitrogen blanket to avoid oxidation of the cuprous ions present. Finally, the cuprous content was calculated as the difference between total soluble copper and cupric ion.

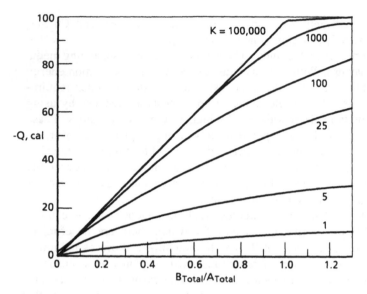

Figure 12   Calorimetric titration of 0.1F A with B $\Delta H°$ for the reaction is -10 kcal/mol in all cases.

APPENDIX II: CALORIMETRIC DETERMINATION OF EQUILIBRIUM CONSTANTS

The shape of a thermometric titration curve is a function of both the log K and $\Delta H°$ values for the reaction taking place during the titration. If the log K value for the reaction is small enough that the reaction can be considered reversible, the thermometric titration curve associated with the reaction will be nonlinear and the curvature will depend on the magnitude of the log K value for the reaction. This is illustrated in Figure 12, where idealized titration curves are given for several reactions having the same $\Delta H°$ but differing log K values. It is then possible to analyze the data and calculate both log K and $\Delta H°$ values for the reaction occurring in the calorimeter.

The method used for analysis of the calorimetric data was described previously [38,39].

APPENDIX III: THERMAL STABILITY OF CUPROUS TRIFLUOROACETATE AND CUPROUS NITRATE SOLUTIONS

Thermal Stability of CuTFA in Concentrated Xylene and Propionitrile Solutions

Thermal stability of CuTFA/Xylene solutions was studied at 100, 135, 160, and 200°C. At 200°C decomposition proceeds by the reaction:

$$\text{CuOOCCF}_3 + \underset{\text{CH}_3}{\overset{\text{CH}_3}{\bigodot}} \rightarrow \text{Cu}^0 + \text{HOOCCF}_3 + 1/2 \left(\underset{\text{CH}_2}{\overset{\text{CH}_3}{\bigodot}} \right)_2$$

+ other aromatics   (III-1)

No decarboxylation was detected.

Accumulation of HTFA accelerates the reaction. At 135° and 160°C the rate of reaction is second order in [CuTFA] and first order in [HTFA].

$$\frac{-d[\text{CuTFA}]}{dt} = K_{21}[\text{CuTFA}]^2[\text{HTFA}] = \text{``}K_1\text{''}[\text{CuTFA}] \qquad (111\text{-}2)$$

rate constants at 110°C and 125°C were $0.6 \pm 0.3 \times 10^{-4}$ hr^{-1} and $4.7 \pm 1 \times 10^{-4}$ hr^{-1}. Comparisons with the rate constants for decomposition of CuTFA/xylene indicate that the 1:3 CuTFA/EtCN solvents are more stable at 100°C, while 1:1 CuTFA/xylene solvents become more stable at ~120°C. By extrapolating the 125 and 110°C data, we obtain a half-life of 1000 days at 100°C for the CuTFA/3EtCN.

The decomposition of CuTFA in EtCN at 125°C proceeds predominantly by decarboxylation.

$$\text{CuOOCCF}_3 \rightarrow \text{Cu}^0 + \text{CO}_2 + \text{CF}_3{}^\cdot \qquad (III\text{-}3)$$

$$2\text{CF}_3{}^\cdot \rightarrow \text{C}_2\text{F}_6 \qquad (III\text{-}4)$$

The fact that Kolbe-type decarboxylation is possible in propionitrile but not in xylenes is probably due to the electron transfer abilities of the nitrile. According to the literature, the trifluoroacetate anion is stable at temperatures below 200°C [40].

Thermal Stability of CuNO₃ in Propionitrile Solutions

Decomposition rates were determined from the rate of increase in pressure caused by NO$_2$ generated by decomposition of CuNO$_3$/EtCN in closed vessels. The decomposition reaction is:

$$\text{CuNO}_3 \rightarrow \text{CuO} + \text{NO}_2 \qquad (III\text{-}5)$$

At the temperature and pressure ranges investigated (95–115°C and 1–15 atm of NO$_2$) less than 20% of the NO$_2$ associates to N$_2$O$_4$ [41]. The extent of association was taken into consideration in interpreting the decomposition rates.

Decomposition rates were studied for a CuNO$_3$/EtCN solvent containing 17.4 wt % Cu(I). The experiment lasted 22 days. At 95°C, the lifetime of this system, assuming the solvent composition is kept constant by continuous makeup and bleed, is 170 days. In a process the solvent is at elevated temperatures only in the stripper reboiler, i.e., only for a fraction of the time. For a typical extractive distillation of ethylene and propylene from the C$_1$–C$_3$ fraction, a 2.5-min residence in the reboiler leads to a solvent lifetime of 1300 days at 95°C.

An increase in concentration leads to higher rates of decomposition. To compensate for the increased decomposition rate we must reduce the maximum operating (reboiler) temperature. An increase in concentration from 17 wt % Cu(I) to 22 wt % Cu(I) can be compensated by a reduction in maximum temperature of about 5°C, i.e., 1°C per wt % Cu(I) increase.

APPENDIX IV: PHASE DIAGRAM DETERMINATIONS

Ternary phase diagrams were constructed from analytical (GLC) determination of the composition of the polar and nonpolar phases in equilibrium. The samples, $15-25$ cm^3 in volume, were equilibrated under an N_2 blanket. Equilibration was usually completed within a few minutes. A convenient experimental procedure was to prepare the CuTFA/EtCN solution under N_2 in a dry box, place it in a vial, add the appropriate paraffin, and cover the vial with a serum cup. The vial and taken out of the dry box, and analysis of the upper and lower phase gave the binary compositions for the base of the ternary diagram (EtCN-paraffin side). Olefin was then added to obtain tie lines corresponding to higher solvent loadings. For the same ternary, measurements could also be started at the EtCN-olefin side of the triangle. In this case paraffin was added to provide the lines with increasingly lower loading. Usually a ternary diagram was based on complementary tie line data obtained from two different samples. The CuTFA in the lower phase did not interfere with the analysis because the complex would break up in the hot injector port of the GLC unit.

To determine distribution coefficients at 50°C, an aqueous thermostated bath was used. The ternary diagrams (shown in Figures 5–10) were constructed on a CuTFA-free basis, i.e., on the basis of volatile components alone.

REFERENCES

1. W. C. Zeise, *Mag. Pharm. 35*, 15, (1830).
2. F. R. Hartley, *Chem. Rev. 69*, 799, (1969).
3. M. J. S. Dewar, *Bull. Soc. Chim. France 18*, C79, (1951).
4. J. Chatt and L. A. Duncanson, *J. Chem. Soc.* 2939 (1953).
5. R. G. Solomon and J. K. Kochi, *J. Am. Chem. Soc. 95*(6), 1889(1973).
6. C. E. Morrell et al., Trans, *A.I.Ch.E. 42*, 473 (1946).
7. R. B. Long, *Recent Developments in Separations Science*, Vol. 1 (N. Li, ed.). CRC Press, Cleveland, Ohio, 1973.
8. M. A. Muhs and F. T. Weiss, *J. Am. Chem. Soc. 89*, 4697 (1962).
9. K. N. Trueblood and H. J. Lucas, *J. Am. Chem. Soc. 74*, 1338 (1952).
10. I. V. Nelson et al., *J. Inorg. Nucl. Chem. 22*(3/4), 279 (1961).
11. R. G. Pearson, *Science 151*, 172 (1966).
12. G. J. Pierotti, C. H. Deal, and E. L. Derr, *Ind. E. Chem. 51*, 95 (1959).
13. D. A. McCayley, U.S. Patent 2,953,589 (Sept. 10, 1960).

14. R. W. Turner and E. L. Amma, *J. Chem. Soc. 85*, 4046 (1963).
15. A. K. Dunlop, G. C. Blytas, and E. R. Bell, U.S. Patent 3,401,112 (September 10, 1968).
16. G. C. Blytas, U.S. Patent, 3,546,106 (Dec. 8, 1970).
17. G. C. Blytas, U.S. Patent, 3,801,666 (April 2, 1974).
18. G. C. Blytas, U.S. Patent, 3,801,664 (April 9, 1974).
19. G. C. Blytas, U.S. Patent, 3,520,947 (July 21, 1970).
20. G. C. Blytas, E. R. Bell, and A. K. Dunlop, U.S. Patent 3,449,240 (June 10, 1969).
21. R. B. Long, H. H. Horowitz, and D. W. Savage, U.S. Patent 3,754,047 (Aug. 21, 1973).
22. R. B. Long, F. A. Caruso, J. P. Longwell, and R. J. DeFeo (U.S. Patent 3,592,865).
23. P. J. Haase, P. M. Duke, and J. W. Cates, *Hydroc. Processing* (March 1982).
24. D. J. Haase and D. G. Walter, *Chem. Eng. Progr. 70*(5) (May 1974).
25. A. P. Guiterrez et al., ESEP®, A Process for the Recovery of Ethylene, 175th ACS Meeting, Anaheim, CA, March 12–17, 1978.
26. E. R. Gilliland et al., *J. Am. Chem. Soc. 61*, 1960 (1939).
27. H. W. Quinn and D. N. Glew, *Can. J. Chem. 40*, 1103 (1962).
28. A. L. Ward and E. C. Makin, Jr., *J. Am. Chem. Soc. 69*, 657 (1947).
29. J. Kendall, E. D. Crittenden, and H. K. Miller, *J. Am. Chem. Soc. 45*, 963 (1923).
30. J. J. Christensen, R. M. Izatt, L. D. Hansen, and J. H. Partridge, *J. Phys. Chem. 70*, 2003 (1966).
31. F. Becker, J. Barthal, N. G. Schmahl, and H. M. Lucow, *Z. Phys. Chem. 37*, 52 (1963).
32. R. M. Izatt, D. J. Eatough, J. J. Christensen, and R. L. Snow, *J. Phys. Chem. 72*, 1208 (1968).
33. J. C. Frank, G. R. Geyer, and H. Kehde, *Chem. Eng. Progr. 65*(2), 79 (1969).
34. D. C. Tabler and M. M. Johnson, U.S. Patent 4,129,605 (Dec. 12, 1978).
35. T. Sato et al., *J. Chem. Eng. Jpn. 21*(2), 192 (1988).
36. G. C. Blytas, U.S. Patent 3,415,615 (Dec. 10, 1968).
37. G. C. Blytas, U.S. Patent 3,656,886 (April 18, 1972).
38. P. G. Blake and H. Pritchard, *J. Chem. Soc. 13*(4), 282 (1967).
39. J. J. Christensen, D. J. Eatough, J. Ruckman, and R. M. Izatt, *Thermoclimica Acta 3*, 203 (1972).
40. D. J. Eatough, E. M. Izatt, and J. J. Christensen, *Thermoclimica Acta 3*, 233 (1972).
41. F. Daniels and R. Alberty, *Physical Chemistry*. Wiley, New York, London, 1955, p. 241.

3

Olefin Recovery and Purification via Silver Complexation

George E. Keller, Arthur E. Marcinkowsky, Surendra K. Verma, and Kenneth Dale Williamson *Union Carbide Chemicals & Plastics Co., Inc., South Charleston, West Virginia*

INTRODUCTION

Cryogenic distillation has been used for over 60 years [1] for the recovery and refining of ethylene and propylene from olefin plants, refinery gas streams, and other sources. Today this technology stands virtually unchallenged in this service. Figure 1 is a schematic diagram of a typical cryogenic-based process, and Table 1 gives a typical analysis of a feed to the refining system. As can be seen in Figure 1, parts of the process operate at considerably below ambient temperature and considerably above atmospheric pressure. Thus ethylene and propylene recovery and purification can be costly in terms of both investment and energy costs.

In recognition of these costs, several attempts have been made to develop different recovery approaches, most of which would obviate the need for sub-ambient temperatures. The most popular approach has been π-bond complexing, in which either a silver or a cuprous ion bonds selectivity with the olefin. Different counterions and solvents [2] have been used to influence the tenacity of these bonds. The literature is not voluminous, however, and there is a need for further quantitative information so that the true worth of π-bond complexing as a means of olefin recovery and purification can be assessed.

The objective of this chapter is to present a study of one system: aqueous silver nitrate. Following a description of the existing cryogenic process and a discussion of π-bond chemistry, we present a vapor–liquid equilibrium package for ethylene and propylene over a range of temperature and silver nitrate molarity, as well as other relevant data. We also present other process chemistry aspects, an

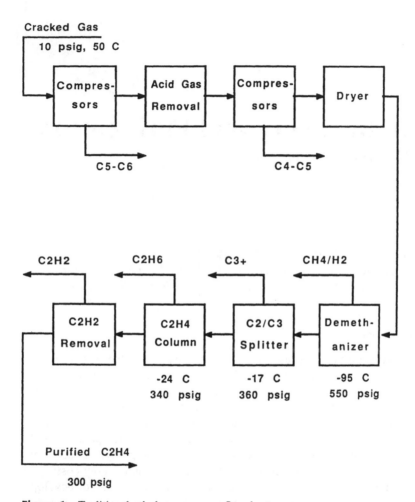

Figure 1 Traditional ethylene recovery flowsheet.

engineering analysis of the process, and an analysis of the concept of satisfactory risk in commercializing a new process.

CONVENTIONAL OLEFIN RECOVERY SYSTEM

A typical conventional olefin production process cracks a hydrocarbon feed containing ethane or higher molecular weight feeds up to gas oil in high-temperature pyrolysis furnaces. The cracked gas is then cooled from about 750–950°C at the furnace outlet to about 400°C in a heat exchanger in which high-

Table 1 Typical Analysis of Cracked Gas
Stream to Refining Unit

Component	Composition (mol %)
Hydrogen	15.8
Nitrogen	0.2
Carbon monoxide	0.3
Methane	26.1
Acetylene	0.2
Ethylene	25.5
Ethane	3.8
Methyl acetylene/propadiene	0.2
Propylene	10.8
Propane	15.1
1,3-Butadiene	0.7
Butenes/butanes	0.5
C_5's	0.4
C_6 nonaromatics	0.1
Benzene	0.2
C_{7+} hydrocarbons	0.1
	100.0

pressure steam is generated. Before entering the first separation stage, the furnace gas stream is further quenched to 40°C or less to remove high-boiling components and leave a gas stream consisting primarily of C_4 and lighter components. This stream is subsequently compressed to about 500–600 psig in several stages with intermediate cooling. In these coolers small amounts of C_5 and heavier components and some water condense and leave the system. Amine-based and caustic-based scrubbers are used to remove acid gases (primarily carbon dioxide and hydrogen sulfide), and molecular sieve-based dryers are used to remove the remaining water vapor before subjecting the gas to cryogenic conditions. Removal of water vapor and carbon dioxide is essential to prevent their freezing in the cryogenic columns and plugging them. Hydrogen sulfide must be removed to prevent its contaminating the light olefin products. Acetylenes can also be removed between two stages in the compression train or later in the process. Removal can be effected by either hydrogenation or absorption into acetone or another polar solvent.

Generally the first cryogenic step consists of removing the fuel gases (methane and hydrogen) from the stream via distillation. Several different schemes can then be used to separate the C_2–C_4 stream into its individual components. One scheme—the one shown in Figure 1—is to split this stream into a C_2 and a C_{3+}

stream. Acetylene, if not removed previously, can be removed at this place in the process. Ethylene is then purified by removing the ethane, which in turn is recycled to the furnaces for cracking. Ethane-ethylene separation is difficult to accomplish by distillation and is carried out at about $-25°C$ and 320 psig in a column containing over 100 trays. The remaining C_3-C_4 stream is sent to a C_3/C_4 splitter. Propylene is also separated from propane by a very energy-intensive distillation at about $-30°C$ and 30 psig.

π-BOND COMPLEXATION

Metal ion–olefin complexation or, in a more general sense, π-bond complexation and its reversibility is today a well-known physicochemical phenomenon [3]. Historically, the first recognized metal ion–olefin complex was the platinum-ethylene complex prepared by Zeise in the early 1830s [4]. Reversibility and the potential for industrial scale separation and purification was not addressed until early in the 20th century, stemming from the work of Bertholet and later Manchat and Brandt [5], who reported on the reversible absorption/desorption nature of the Cu(I)-ethylene system.

Presumably the first successful industrial process using metal ion complexation was the extraction of carbon monoxide by cuprous salts introduced by Bosch as part of the technology for ammonia synthesis [6]. The earliest successful pilot plant scale efforts to recover ethylene are attributed to Horsley [7,8], who used concentrated aqueous $AgNO_3$ solutions.

In general, the metal ions most effective in forming reversible π-bond complexes which can be used in large-scale separation and purification processes are Ag(I) and Cu(I); other cations known to form complexes are Pt(II), Pd(II), and Hg(II). Silver- and copper-based complexes are easily reversible using temperature and/or pressure swings. Those of the latter group are more strongly bonding and difficult to reverse without destroying, i.e., converting to another product, the original olefin. The use of sophisticated ligand stabilization techniques could conceivably transform other cations into truly reversible systems, but such chemistry is beyond the scope of this paper.

A review of the early hypotheses concerning metal-olefin bonding, concluding with the currently held donor–acceptor bonding model developed by Dewar in 1951, is given by Herberhold [3,9]. Dewar's theories were subsequently generalized by Chatt and Duncanson [9]. This Dewar–Chatt model of the metal-olefin bond, which has been accepted for three decades, is generally referred to as π-bonding. But it should be noted that in using this abbreviated notation, there is nevertheless always s-type bonding present which can vary depending on the specific metal and olefin involved in the complex. Thus, according to Murrell and Scollary [10], the important role of s orbitals in the bonding of Ag^+ provides an explanation for the fact that the $Ag(I)-C_2F_4$ complex is more stable than Ag(I)-

C_2H_4 complex. It has also been reported that the Ag(I)-deuterated C_2H_4 complex is stronger than the Ag(I)-C_2H_4 species [11] which is likewise in accord with the Dewar–Chatt model. The model nevertheless is powerless to deal with most of the "engineering" variables, such as the nature of the silver salt anion, silver salt concentration, type of solvent, temperature, pressure, and added electrolyte—all of which can have a pronounced effect.

An excellent review by Quinn [2], dealing with the nature of the silver-olefin complex, absorption differences in aqueous $AgNO_3$ and silver fluoborate ($AgBF_4$) solutions, chromatographic separations, anhydrous salts, and silver-exchanged resins and zeolites, was published in 1971.

In general, the solubility of olefins in aqueous silver solutions is dependent on the anion of the salt and the salt concentration. Below 1 M, olefin solubility per mol of silver decreases in the order $ClO_4 >> BF_4 > NO_3$ for the respective salts, whereas in more concentrated solutions the olefin solubility decreases in the order $BF_4 >> ClO_4 > NO_3$, which parallels the strengths of the corresponding acids. For ethylene Baker [12] observed an increase in its solubility per mol of Ag in $AgNO_3$ solutions with increasing electrolyte concentration. The addition of fluoborates including HBF_4 increases the solubility of ethylene in both aqueous $AgBF_4$ and $AgNO_3$ solutions while for $AgNO_3$ solutions the addition of HNO_3 increases the ethylene solubility, but addition of $NaNO_3$ depresses its solubility [13,14]. Wilcox and Goal [15] point out that alkyl substitution on ethylene lowers the Ag-olefin complex strength, resulting primarily from steric hindrance.

A potential problem with π-complexing metal ions is that they can form irreversible complexes with acetylenes, especially Cu(I) and Ag(I). These complexes can be highly unstable and susceptible to detonation, especially when they build up in a system to the extent that they precipitate out of solution. These precipitates become even more sensitive to detonation once they become dry, and hence elaborate precautions must be taken to deal with them effectively on a large scale.

All aqueous solutions of Ag(I), including $AgNO_3$, are stable in the presence of saturated hydrocarbons, carbon monoxide, and carbon dioxide. In the presence of hydrogen, which is always present in crude feed streams, a gradual reduction to metallic silver is observed [2]. This reduction problem must be eliminated in order to have a commercially successful purification system. If this cannot be done, then silver will be continually lost through colloidal particle formation and through plating out on the walls of vessels and pipes.

Acetylene reacts with Ag(I) regardless of the nature of the anions present. The solubility of silver acetylide (Ag_2C_2) in $AgBF_4$ appears to be greater than in $AgNO_3$ [16].

In aqueous $AgNO_3$, two acetylide species are formed [16]. One is $Ag_2C_2 \cdot AgNO_3$, which is formed in solutions with $AgNO_3$ concentrations less than 10 wt %. Above 25 wt % $AgNO_3$, $Ag_2C_2 \cdot 7AgNO_3$ is formed. Shaw and Fisher also state that Ag_2C_2 is very soluble in strong $AgNO_3$ solutions. These complexed acetylide species are not as shock-sensitive as the uncomplexed Ag_2C_2. Ag_2C_2 decomposition studies were reported by McCowan [17].

COMMERCIAL AND PROPOSED PROCESSES

Only a very few π-bond-based olefin recovery processes have ever been commercialized. The use of cuprous nitrate–ethanolamine solutions for ethylene recovery was developed and practiced in Germany during World War II. This process is no longer in operation. Hoechst in West Germany developed a process in the late 1950s for ethylene recovery based on a concentrated aqueous solution of $AgBF_4$ and fluoboric acid [18–20]. This process was taken through the pilot plant and demonstration stages, but it was apparently not taken to full commercialization. During development of this process, a very thorough engineering effort was expended, and workable solutions for both the acetylide and silver ion reduction problems were achieved. More recently, Tenneco using basic information developed by Exxon developed a process, called Cosorb, for recovery of carbon monoxide using a nonaqueous cuprous aluminum chloride–aromatic solvent solution [21–23]. An extension of this process for the recovery of ethylene has been advertised [24] but it is apparently not practiced commercially.

Many other π-bonding processes have been proposed. These include the recovery of other olefins, the use of various anions, the use of solvents other than water, the use of salts and acids as additives, the use of alternative modes of contacting such as solvent extraction, and the use of membrane-based processes involving π-complexes in the membrane to form a pathway for facilitated transport for the olefin. A few examples follow.

1. Separation of gaseous olefins from paraffins by complexing with solid Cu(I) salts [25].
2. Membrane processes for ethylene recovery utilizing metal ions such as Ag(I) and Cu(I) contained in a so-called liquid barrier held in place by permeable membranes [26–29].
3. Use of concentrated $AgBF_4$ and $AgSiF_6$ for the primary recovery of styrene from C_8 hydrocarbons [30,31].
4. Styrene separation from ethylbenzene via liquid–liquid extraction using concentrated cuprous nitrate–propionitrile in combination with a C_5–C_{18} paraffin solution [32].
5. Recovery of C_6–C_{10} olefins from paraffins using Ag(I) or Cu(I) in alkanol solutions [33].
6. Hydrocarbon separation using liquid exchange reactions involving $CuBF_4$ stabilized with various coordinating compounds [34].
7. Use of silver phenylsulfonate in water to extract pentene from isopentane [35].
8. Use of aqueous silver hexafluorophosphate to recover olefins [36].

9. Separation of styrene and ethylbenzene via facilitated transport through a perfluorosulfonate ionomer membrane exchanged with silver ions [37].

Many other examples of proposed processes exist. But in spite of the wide variety of processes investigated and the great amount of ingenuity expended, there is no indication that any has been commercialized to date.

EXPERIMENTAL DATA AND MODELING

In this section we present a package of phase equilibrium and other data associated with a process for recovery of ethylene from a prepurified cracked-gas stream.

Experimental Procedures

The equilibrium amount of ethylene absorbed by aqueous $AgNO_3$ solution was determined by two procedures. For pressures near atmospheric, a volumetric scheme was used. Ethylene taken from a reservoir into a gas burette was passed into a cell containing a measured volume of silver nitrate solution at a given concentration. A magnetic stirrer hastened the attainment of equilibrium with the vapor phase. Use of a leveling bulb permitted the adjustment of the pressure to the desired level.

The second procedure was used to establish equilibrium at higher partial pressures of ethylene—up to about 150 psia. Absorption at temperatures from 5 to 60°C and concentrations of $AgNO_3$ from 3 to 8 M (as measured at 25°C) were included in the study. The cell used was a tensimeter, a stirred stainless steel bomb of about 390-ml capacity. Measured amounts of solution were added to the cell, which was then closed off, and pumped briefly, to remove air. (A very small amount of hydrogen peroxide was added to the $AgNO_3$ solution prior to loading into the cell to prevent Ag(I) reduction to metallic silver). Ethylene was then added to the cell from a weighed, high-pressure bomb. The amount of ethylene added was established by reweighing the bomb. The net weight less the calculated weight of the ethylene in the vapor phase was the weight of the absorbed ethylene. A calibrated Heise gage gave the pressure at which the absorption occurred. Successive additions of ethylene were made in this matter, to establish the relationship of pressure vs. concentration at a given temperature.

The region of greatest interest for an ethylene separation process is covered by temperatures from 0 to 60°C, pressures from 0 to about 150 psia, and concentrations of 3 to 8 M $AgNO_3$. The data reported here are within those limits.

Aqueous $AgNO_3$ has been characterized fairly completely (see, e.g., Ref. 38); the solution loaded with ethylene has been characterized hardly at all. Addition of $AgNO_3$ to water depresses the freezing point to a minimum of about $-6°C$. The viscosity increases by more than a factor of 2 in going from pure water to 7.7 M $AgNO_3$. The surface tension rises only modestly from that of pure water to concentrated $AgNO_3$ solutions. The solution is a strong electrolyte. At 18°C and

0.01 M, α, the fraction dissociated, is 0.9963. However, α decreases appreciably at higher concentrations. At 25°C and 1 M, α is only 0.714, while at 9.7 M, α = 0.239. The expression

$$\alpha = \frac{[Ag(I)][NO_3]}{[AgNO_3]}$$

is surely an oversimplification. Cryoscopic measurements and observations of the Raman spectra indicate that complexes such as $[Ag(NO_3)_2]^-$ and $[Ag_2(NO_3)]^+$ are present in concentrated solutions.

The vapor pressure of the water is suppressed by the addition of $AgNO_3$. Up to quite high concentrations, the effect is close to what is calculated on the basis of the added solute, as ionized. Solubilities of $AgNO_3$ are very large. For example, at 35°C saturation is reached at 10.9 M (75.4 wt %).

Phase Equilibrium Data

The absorption of ethylene by $AgNO_3$ solutions is shown in Table 2 and in Figures 2–6. The data are represented in these figures as R vs. ethylene pressure (with R = mol ethylene absorbed per mol of $AgNO_3$ in solution). Data for each temperature are presented on a separate figure. The 8 M $AgNO_3$ data are not depicted because they are available for only two temperatures.

To check the internal consistency of the data, the experimental R vs. pressure curves were cross-plotted in a variety of ways. To get discrete values of pressure for these purposes, faired curves were generated. Values of R at given temperatures and molarities were taken from these curves at 50, 100, and 150 psia, for various types of cross-plots. Faired values taken from these plots provided values for the solid lines in the figures; these lines are felt to be more reliable than the individual points.

Qualitatively the system behaves as follows: At a given temperature and pressure, the higher the concentration of $AgNO_3$, the greater the amount of ethylene absorbed (though the higher the concentration of solution, the lower the ratio of ethylene to $AgNO_3$ at equilibrium). Absorption is a strong function of temperature, i.e., the lower the temperature, the greater the absorption. Also, for a given $AgNO_3$ concentration, there is a linearly increasing absorption with increasing pressure at low pressures. At higher pressures, R appears to approach a limit. This behavior comes closer to following a Langmuir model than to following Henry's law. However, R ultimately can become greater than unity. This may indicate an equilibrium with two or more complex species. It should also be remembered that not all of the $AgNO_3$ is ionized, so that if the value of R were calculated based on just the ionized $AgNO_3$, it could be considerably higher.

There is a small amount of ordinary solution of ethylene in the liquid phase. Discerning the amount of simple solution is not easy, because this probably small effect is masked by the larger effect of complexing. Moreover, the solution takes place in a concentrated electrolytic solution, a field not yet covered very ade-

Table 2 Ethylene Absorption in Aqueous Silver Nitrate

Temperature (°C)	Molarity	C_2H_4 pressure (psia)	Mol C_2H_4/ Mol $AgNO_3$=R
5.0	3.00	13.0	0.255
		21.3	0.497
		32.4	0.696
		50.2	0.856
		78.6	1.013
		112.2	1.122
		143.5	1.221
	4.00	10.7	0.230
		21.0	0.419
		29.7	0.584
		42.0	0.711
		53.8	0.861
		70.6	0.967
		88.4	1.060
		104.2	1.143
		125.3	1.235
		151.0	1.321
	6.00	12.6	0.216
		36.8	0.549
		45.1	0.658
		84.5	0.803
		124.1	1.028
25.0	3.00	25.3	0.457
		49.5	0.671
		74.5	0.798
		103.6	0.915
		122.8	0.971
		148.3	1.027
	6.00	48.8	0.506
		98.3	0.721
		149.9	0.846
		47.5	0.529
		99.0	0.776
		143.7	0.892
		21.7	0.261
		33.6	0.346
		43.7	0.442
		60.0	0.538
		72.7	0.598
		89.5	0.671
		105.2	0.737
		122.5	0.786
		151.8	0.856

Table 2 Continued

Temperature (°C)	Molarity	C_2H_4 pressure (psia)	Mol C_2H_4/ Mol $AgNO_3$=R
	8.00	23.1	0.251
		48.7	0.605
		72.8	0.743
		106.2	0.844
		127.0	0.898
		149.9	0.957
		177.3	1.022
		206.2	1.076
40.0	4.00	25.8	0.251
		48.1	0.411
		76.0	0.529
		101.3	0.617
		124.7	0.678
		151.7	0.988
	6.00	21.3	0.139
		50.3	0.310
		81.7	0.443
		108.6	0.530
		144.8	0.618
		23.6	0.115
		39.6	0.236
		57.9	0.330
		80.6	0.427
		100.1	0.498
		122.1	0.571
		146.7	0.635
		51.1	0.299
		99.1	0.438
		152.3	0.617
	8.00	24.8	0.155
		50.8	0.290
		72.3	0.424
		99.5	0.518
		120.5	0.582
		150.7	0.662
		178.2	0.716
50.0	3.00	29.6	0.180
		45.5	0.352
		64.6	0.470
		88.3	0.601
		118.3	0.718
		147.8	0.880

Table 2 Continued

Temperature (°C)	Molarity	C_2H_4 pressure (psia)	Mol C_2H_4/ Mol $AgNO_3$=R
	3.00	50.5	0.335
		104.7	0.659
		145.8	0.809
	4.00	36.3	0.203
		53.2	0.329
		74.8	0.458
		98.5	0.559
		122.0	0.632
	6.00	24.8	0.073
		39.7	0.179
		57.8	0.283
		77.5	0.358
		99.0	0.429
		122.0	0.494
		146.0	0.576
60.0	4.00	23.4	0.126
		43.6	0.255
		73.6	0.325
		100.7	0.405
		127.9	0.468
		152.7	0.525
	6.00	39.0	0.144
		68.4	0.253
		86.4	0.300
		105.2	0.354
		121.0	0.393
		142.1	0.436

quately by data or theory. Finally, it is possible that the nature of the solvent changes markedly as the ethylene complex accumulates, giving the solution an organic nature of changed solvent properties. As a result of all these factors, we have found that no simple model, assuming one or several discrete complexes, such as $AgNO_3 \cdot C_2H_4$, $AgNO_3 \cdot 2C_2H_4$, etc., is adequate to fit the experimental data.

In spite of the lack of an adequate equilibrium model, the data show a surprising uniformity in one respect. The heat of complexation is almost invariant, regardless of temperature, pressure or extent of absorption, ranging from 20 to 25 kJ/mol.

Solubility experiments were also made on gases such as methane and carbon monoxide. The latter was especially important to determine since another

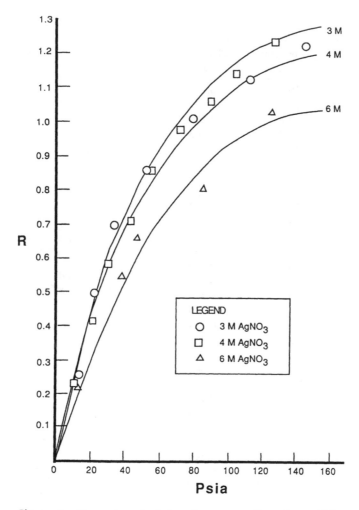

Figure 2   Absorption of ethylene in aqueous silver nitrate, 5°C.

π-bonding ion, Cu(I), complexes carbon monoxide readily, and it was necessary to determine if Ag(I) also had any tendency to do so. There seems to be little or no such tendency with the $AgNO_3$ solution, and carbon monoxide is approximately as soluble as might be expected from physical solubility alone.

SILVER COMPLEXATION-BASED PROCESS

In this section we will describe a process to recover and purify ethylene using an aqueous silver nitrate solution. This process is based on absorption of ethylene into

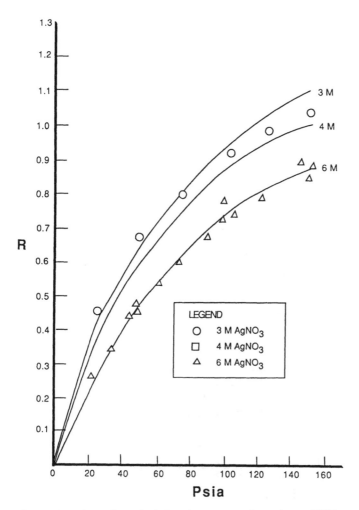

Figure 3 Absorption of ethylene in aqueous silver nitrate, 25°C.

the solution, removal of coabsorbed impurities and subsequent removal of the ethylene. A typical cracked-gas stream contains a number of components which must be removed before the silver complexation process can be employed successfully. If not removed, these components will contaminate the solution and/or the separated ethylene. This prerefining is accomplished in a series of steps, most of which are needed even for the conventional cryogenic process.

All of the heavy components (C_3 and above) in the cracked-gas stream are removed. Fuel oil (C_{10} and above) is first removed in a gasoline fractionator. Dripolene (C_5–C_{10}) is removed primarily in a quench tower. Lighter components

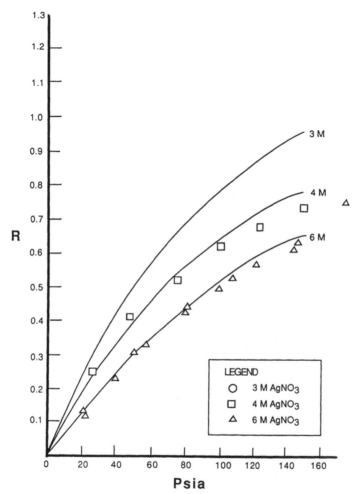

Figure 4 Absorption of ethylene in aqueous silver nitrate, 40°C.

down to C_3s are removed in intercoolers in the compression train and in a C_2/C_{3+} splitter. A conventional acid-gas unit is used to remove hydrogen sulfide and perhaps part of the carbon dioxide. Failure to remove hydrogen sulfide to a higher degree leads to loss of silver ion through formation of a sulfide. Carbon dioxide removal is not important. There is also no need to remove methane, hydrogen, and water vapor.

Acetylene reacts irreversibly with silver ions and forms Ag_2C_2, which not only removes silver ions from solution but also represents a safety hazard when the concentration rises to the point that precipitation occurs. The acetylene concentra-

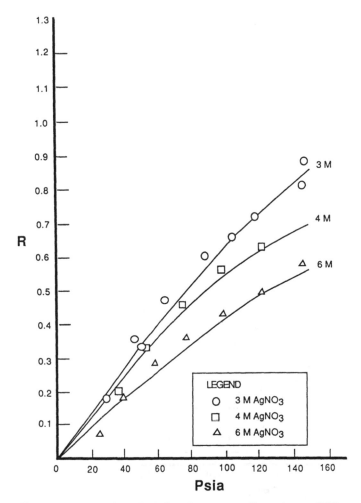

Figure 5   Absorption of ethylene in aqueous silver nitrate, 50°C.

tion in the feed gas can be reduced to below 1 ppm by using either an acetone-based or a dimethylformamide-based absorption process. Even with this small amount of acetylene left in the gas, however, additional means must be taken to prevent Ag_2C_2 buildup in the silver solution; means of controlling this buildup will be discussed later.

Following these prerefining steps, the gas is ready for the silver complexation system. A typical composition of this gas is given in Table 3. A typical product specification for ethylene is also given in Table 4, and the ethylene exiting from this system must meet or improve on all of these values.

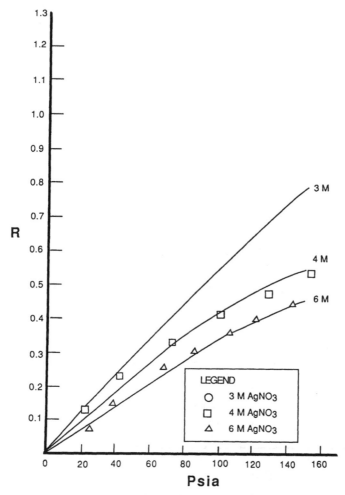

Figure 6   Absorption of ethylene in aqueous silver nitrate, 60°C.

Silver solution stability is a major concern. The presence of hydrogen and perhaps traces of other components has a reducing effect on the silver ions, causing metal precipitation from the solution. This is unacceptable from a process standpoint. We have found [39] that the addition of small amounts of hydrogen peroxide, coupled with the maintenance of a level of nitric acid in the solution stabilizes it against silver precipitation. A synergistic effect occurs by the use of both of these materials such that if the nitric acid is not present, a much greater rate of addition of hydrogen peroxide is necessary to stabilize the solution. In the

Table 3 Typical Feed Gas
Composition to Ethylene
Recovery System

Component	Composition (mol %)
Hydrogen	20.0
Nitrogen	0.3
Carbon monoxide	0.5
Methane	37.0
Acetylene	0.5 ppm
Ethylene	36.5
Ethane	5.6
Propylene	0.1
	100.0

presence of nitric acid, the hydrogen peroxide rate of addition is 1 kg (100%) per 1000 kg of refined ethylene—a completely acceptable rate.

The presence of hydrogen peroxide in the solution causes an in situ oxidation of a small amount of ethylene to carbon monoxide and carbon dioxide. At typical conditions of 3 M silver nitrate, 0.35 wt % hydrogen peroxide and 0.5 wt % nitric acid, the purified ethylene will contain about 30 ppm carbon monoxide and 75 ppm carbon dioxide, as well as about 60 ppm of oxygen from the breakdown of the hydrogen peroxide. These contaminants can be removed, if necessary, with relatively simple scavenging processes downstream of the unit. For example, carbon monoxide and oxygen can be easily removed by copper oxide and metallic copper oxidation, respectively. Carbon dioxide can be removed by a caustic wash, and water can be removed by molecular sieves.

As mentioned above, acetylene reacts instantaneously and irreversibly with

Table 4 Typical Product
Specifications for Ethylene

Component	Maximum level
Methane and ethane	0.15 vol %
Carbon monoxide	5 ppm by vol
Carbon dioxide	10 ppm by vol
Acetylene	10 ppm by vol
Sulfur	5 ppm by vol
Water	10 ppm by vol

silver ion to form silver acetylide (Ag_2C_2). This material in the dry state is highly reactive and readily detonates if subjected to any shock. The solubility limits of silver acetylide in silver nitrate solutions, both with and without ethylene absorbed, were determined, and a safe operating concentration was set at one-half of the lowest value. This concentration can be maintained by the use of silver permanganate as an oxidant [39]. A small sidestream is withdrawn from the stripper and heated to about 75°C under partial vacuum to remove residual ethylene and degrade the hydrogen peroxide. Then solid silver permanganate is added to the solution to destroy the acetylide, forming carbon dioxide and free silver ion. The resulting manganese dioxide precipitates and is filtered out of the solution. The net result is complete removal of the acetylide, complete recovery of the silver, and no introduction of a foreign ion into the solution.

PROCESS ENGINEERING CONSIDERATIONS

The recovery and purification of ethylene in the silver complexation process consists of three distinct steps:

1. Absorption of primarily ethylene from the process gas stream
2. Venting off of various inert impurities from the solution
3. Desorption of the purified ethylene from the solution

A process flow diagram is given in Figure 7, and a listing of the operating conditions and process results is given in Table 5.

Absorption

As indicated in the process flow diagram, the absorption of ethylene is carried out under pressure and in a countercurrent flow absorber. A packed column is specified instead of a trayed column to minimize the inventory of silver solution in the column. The absorption is exothermic, causing the solution to increase in temperature, and it becomes desirable to recover this heat by using it for desorption by means of a heat pump cycle. Removal of the heat of absorption in the absorber, which causes a decrease in temperatures in the absorber, also has the beneficial effect of increasing the ethylene loading in the solution for a given partial pressure of ethylene in the feed.

Venting of Impurities

In the absorption process, small amounts of other components (carbon monoxide, hydrogen, methane, ethane, etc.) of the feed are either physically absorbed in the solution or very weakly complexed with silver ions. The purpose of the venting section is to strip these impurities out of the solution before ethylene is recovered.

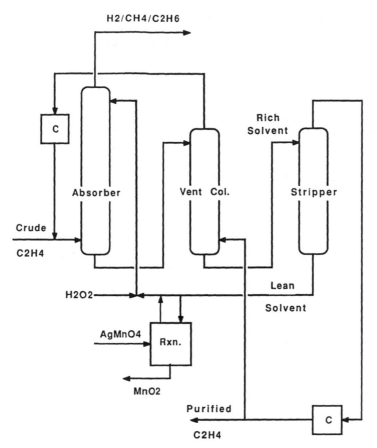

Figure 7   Silver-complexing process.

The vent column operates at a pressure somewhat below the partial pressure of ethylene in the absorber feed. Purging of the impurities is caused by the lower pressure operation and also by feeding a small fraction of the purified ethylene to the bottom of the column as a purge gas. The resulting purge stream contains about 10% of the ethylene in the absorber feed. Such an amount is far too large to be lost from the process, so the stream from the top of the vent column is compressed and added to the absorber feed. After passing through the vent column, the solution contains ethylene at a purity which is generally acceptable for high-purity specifications. As mentioned above, because of the presence of hydrogen peroxide in the solution, very small amounts of oxygen and carbon oxides will be present in

Table 5 Operating Conditions and Process Results

Conditions	AgNO$_3$ molarity	
	3	6
Absorber		
Feed total pressure, psia	240	240
Ethylene partial pressure in feed gas, psia	95	95
Absorber top temperature, °C	30	30
Absorber base temperature, °C	40	40
R at absorber base	0.60	0.395
Ethylene absorbed in the absorber, mol/liter solution	2.18	2.37
Ethylene mole fraction in absorber overhead	0.015	0.015
Stripper		
Total pressure, mm Hg abs.	450	450
Ethylene partial pressure, mm Hg	364	370
Water vapor partial pressure, mm Hg	86	80
Temperature, °C	50	50
R at stripper base	0.046	0.033

Note: Absorber feed includes the vent column overhead. This additional stream consists of a flow of about 10% of the ethylene absorbed in the absorber plus smaller flows of impurities stripped in the vent column.

the ethylene product, and for some applications these levels may have to be removed by scavenging techniques.

Desorption

Desorption or stripping of ethylene can be carried out either in one stage or in more, with each subsequent stage operating at a successively lower pressure. More stages would reduce the energy consumption by reducing the compression requirement for raising the ethylene pressure from the higher pressure stages to the delivery value. Capital costs, however, would also increase for a multistage case. A substantial amount of the heat required for desorption is supplied by heat pumping of the heat of absorption over into the strippers.

The final desorption stage is operated at subatmospheric pressure to reach a high degree of recovery of ethylene contained in the solution. Of course, the lower the concentration of ethylene in the solution fed to the absorber, the lower will be the concentration of ethylene lost in the gas leaving the absorber. In our pilot plant a one-stage stripper was operated at nine psia and 50°C. These conditions produced a 98% recovery of all of the ethylene in the feed to the absorber.

ECONOMICS AND PROCESS RISK

An economic analysis of the silver complexation process has been made for an olefin plant producing 1 billion pounds of ethylene per year. This analysis showed that the savings resulting from the use of this process compared to a highly optimized, conventional cryogenic process was in excess of $1 million per year. This savings resulted both from a lower investment (which reduced the capital cost) and from a reduced energy cost. Pilot plant information showed that the process is quite stable and easily controllable with respect to maintaining the solution active and safe with respect to maintaining low levels of silver acetylide.

Nevertheless, there has been a reluctance to commercialize the process, and our lack of success in "selling" it has led us to study the problem of the risk associated with commercializing a radically new technology. In the context of a billion-pounds-per-year ethylene plant using naphtha as the feed, the million-plus dollars per year savings is about the sales value of one day's product sales. Thus if the new process caused an extra few days of downtime per year, or if the plant startup were delayed by a month or so (as might be expected with a new process), then the projected savings would be swallowed up. In addition, there was a concern about the loss of solvent, which in the case of a silver solution is far from negligible.

We have concluded that at least three questions can be asked regarding the issue of profitability for a new technology.

1. Just how profitable must this technology be with respect to an older, well-proven technology be to be accepted as a replacement?
2. What part does the perceived risk play in the new technology's acceptance for commercialization?
3. How does the required profit improvement vary with the process's product value?

Answers to these questions would surely help in providing guidelines for introducing new technologies.

Our preliminary answers to these questions are shown first in Figure 8. Total yearly savings vs. the sales income of the product per day are plotted on a logarithmic scale. The figure shows two response regions. For relatively low sales incomes, the required savings is nearly the same. In this region, a new process must exhibit enough savings to justify the cost of its development plus other costs that might accrue. Certainly a proposed new process which would save a manufacturer only about $1000 per year would not likely be commercialized if the existing technology performed adequately. The risk would simply not be worth it. We project a threshold savings of about $50,000–$100,000 per year to be required for a new technology to be commercialized even on a very small scale.

As the rate of income increases, a point is reached at which the projected

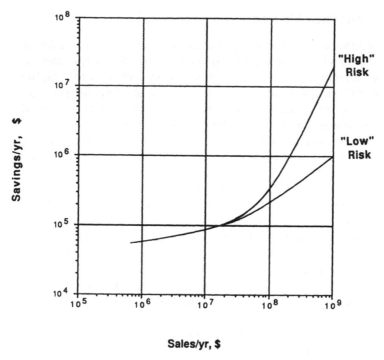

Figure 8   Risk–reward analysis.

savings rises linearly or even more rapidly with income. Also, instead of the trend being a single line, a band results, with an additional parameter of perceived risk associated with the new technology. This risk perception parameter is projected to be quite important, though somewhat difficult to quantify.

Our analysis shows how hard it can be to commercialize new technology on a very large scale, but it also suggests that there can be a pathway to reaching that scale. The strategy is to reduce the perceived process risk associated with large scale. This pathway is shown in Figure 9. By commercializing the technology on small scale (sometimes even when the profitability is not enough for that scale), problems can be worked out and the perceived risk reduced. The small unit can even be thought of as a "commercial pilot plant." This strategy will then allow larger embodiments of the technology to be commercialized without having to meet such a prohibitively large savings level.

The savings values suggested in this analysis are clearly only semiquantitative, and more case studies will be required to improve the quantitative nature of the curves. Nevertheless, the projected trends seem to have some validity and

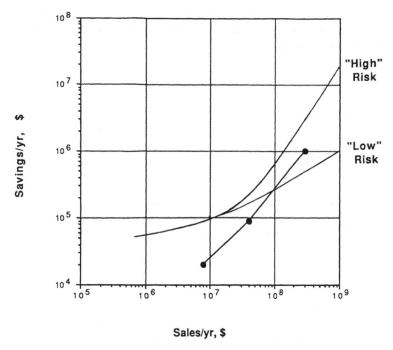

Figure 9   Risk–reward analysis: A strategy.

should serve as a basis of further study of risk and reward in new process introduction.

There may be another parameter to be dealt with in this analysis, however. Different companies may have different perceptions of risk, and as a result the position of the band may be different for them. Such an analysis is surely beyond the scope of this paper.

CONCLUSIONS

An ethylene recovery and refining process based on absorption with an aqueous solution of $AgNO_3$ appears to be technically feasible in all aspects. A reasonable flowsheet results. Problems such as maintaining solution stability, controlling the Ag_2C_2 level at a nonhazardous value in the solution, and producing an acceptably pure product at high recovery have been solved. Significant savings—on the order of a million dollars per year for a billion-pounds-per-year ethylene plant—seem to result compared to the traditional cryogenic-based process. An analysis of process risk has led us to believe that the most likely path for commercializing the silver-complexing-based process for ethylene production (as well as any other process on large scale) will involve its prior commercialization on a much smaller scale. Only in such a way can process risks be minimized to the extent that they will be acceptable for the large-scale plant.

REFERENCES

1. A. E. Marcinkowsky and G. E. Keller, Ethylene and its derivatives: Their chemical engineering genesis and evolution at Union Carbide Corporation. In *A Century of Chemical Engineering* (W. F. Fowler, ed.), Plenum Press, New York, 1982.
2. H. W. Quinn, *Progress in Separation and Purification*, Vol. 4 (E. S. Perry, ed.), Interscience, New York, 1971, p. 133.
3. M. Herberhold, *Metal Π Complexes*, Vol. 2, Part 1, Elsevier, New York, 1972.
4. L. B. Hunt, The first organo-metallic compounds. *Platinum Metals Review*, Vol. 28, Johnson Mathey, London, 1984, p. 76.
5. S. A. Miller, *Ethylene and Its Industrial Derivatives*, 18, Ernest Benn, Ltd., London, 1969.
6. S. A. Miller, *Ethylene and Its Industrial Derivatives*, 96, Ernest Benn, Ltd., London, 1969.
7. British Patent 291,186 (1928).
8. U.S. Patent, 1,875,290 (1932).
9. M. Herberhold, *Metal Π Complexes*, Vol. 2, Part 2, Elsevier, New York, 1974.
10. J. N. Murrell and C. E. Scollary, *J. Chem. Soc. Dalton Trans.*, 1034 (1977).
11. H. J. Emeleus and A. G. Sharpe (eds.), *Advances in Inorganic Chemistry and Radiochemistry*, Vol. 12, p. 144.
12. B. B. Baker, *J. Inorg. Chem. 3*, 200 (1964).
13. J. V. Crooks and A. A. Woallf, *J. Chem. Soc. Dalton Trans. 12*, 1241 (1973).
14. W. Featherstone and A. J. S. Sorrie, *J. Chem. Soc.*, 5253 (1964).
15. C. F. Wilcox and W. J. Goal, *J. Am. Chem. Soc. 93*(10), 2453 (1971).
16. J. A. Shaw and E. Fisher, *J. Am. Chem. Soc. 68*, 2745 (1946).
17. J. D. McCowan, *Trans. Faraday Soc. 59*, 1860 (1963).
18. U.S. Patent, 2,913,505 (1959).
19. H. W. Krekeler, J. M. Hirschbeck and U. Schwenk, *Erdol Kohle 15* 551 (1963).
20. H. W. Krekeler, J. M. Hirschbeck and U. Schwenk, Paper 14, *Proc. Sixth World Petroleum Congress*, Frankfort/Main Sect. IV, June 1963.
21. D. G. Walker, *Chemtech*, 308 (May 1975).
22. *Chem. Week*, 31 (Dec. 15, 1976).
23. U.S. Patent 3,767,725 (1973).
24. *Chem. Eng.*, 18 (May 1979).
25. R. B. Long, Separation of unsaturates by complexing with solid copper salts. In *Recent Developments in Separation Science*, Vol. 1, (N. N. Li, ed.), CRC Press, Cleveland, 1972.
26. U.S. Patent 3,758,603 (1973).
27. U.S. Patent 3,758,605 (1973).
28. U.S. Patent 3,770,842 (1973).
29. U.S. Patent 3,844,735 (1974).

30. U.S. Patent 3,347,948 (1967).
31. W. Featherstone, *Sep. Purif. Meth.* *1*(1), 1 (1972).
32. U.S. Patent 3,801,664 (1974).
33. U.S. Patent 4,132,744 (1979).
34. G. D. Davis and E. C. Makin, Jr., *Sep. Purif. Meth.* *1*(1), 199 (1972).
35. U.S. Patent 2,449,793 (1948).
36. U.S. Patent, 3,189,658 (1965).
37. C. A. Koval, T. Spontarelli, and R. D. Noble, *Ind. Eng. Chem. Res.* *28*(7), 1020 (1989).
38. *Gmelins Handbuch der Anorganishen Chemie*, Verlag Chemie, Weinham, FRG.
39. U.S. Patent 4,174,353 (1979).

4

Immobilized Bioadsorbents for Dissolved Metals

Charles D. Scott   *Oak Ridge National Laboratory,* Oak Ridge, Tennessee*

James N. Petersen   *Washington State University, Pullman, Washington*

INTRODUCTION

It has long been known that many biological materials will accumulate metals. This phenomenon tends to exacerbate the environmental impact of various metals that are released into the environment because they become concentrated in the flora and/or fauna and ultimately get into the food chain. However, these same phenomena potentially can be used in a controlled environment to remove dissolved metals from processing streams either for recovery of valuable resources or to remove pollutants. There is an increasing interest in investigating such biological adsorbents for a variety of applications.

Much of the past research emphasis has been on the investigation of various types of microorganisms for metal adsorption [1–4], but there is also an indication that various plant materials and even animal materials can also be used to accumulate metals [5–7]. Details on the properties of various bioadsorbent materials, mechanisms of interaction, and some specific bioprocessing applications are discussed.

---

*Managed by Martin Marietta Energy Systems, Inc., under contract DE-AC05-840R21400 with the U.S. Department of Energy.

BIOADSORBENT MATERIALS

Biomass that can be employed to remove dissolved metals from aqueous streams has been derived from a variety of sources. A wide variety of different types of biological materials have been studied for this application, at least by scouting tests. In this section, recent work that deals with the ability of microbial biomass to accumulate metals will be examined; then the more limited work on biomass derived from plant and animal materials will be summarized.

Microbial Biomass

Microbial biomass grows rapidly, is relatively inexpensive to generate, and has been demonstrated to have adsorption capacities that are comparable to or can exceed those of inorganic adsorbents [8]. Macaskie and Dean provide an excellent overview of work in this area, summarizing more than 125 publications dealing with adsorption of many different types of metal ions onto various living and nonliving microbial biomass [4].

   Most research in this area has been associated with heavy metals such as Pb, Cd, and Cr. But a broad range of microorganisms has now also been shown to effectively remove metals such as Sr from a typical wastewater containing radioactive pollutants [6,7]. As seen in Table 1, the distribution coefficients for Sr adsorption (quantity of Sr adsorbed per unit mass of dry bioadsorbent divided by the equilibrium concentration of the metal within the bulk solution) can be greater than 80,000 for some microorganisms, although there seems to be great variability depending on type of organisms and exposure times. Operating conditions such as pH, temperature, and so forth, also affect the adsorption properties. Further evaluation of the bioadsorption of Sr by *Microoccus luteus* has shown that bioadsorption accumulation of the metal was a function of the conditions under which the cells were cultured prior to metal loading [9]. Maximum capacity is apparently obtained when the cells are harvested during the latter portion of the exponential growth phase [2,7]. This indicates that the metal-binding sites may be generated as a result of normal cell metabolism. As such, it may be possible to manipulate the growth conditions in order to optimize the generation of these structures. It was also shown [9] that certain physical or chemical pretreatments tend to increase the amount of metal that can be accumulated onto the biomass. Such simple, inexpensive pretreatments may tend to make potential binding sites more accessible to the metal ions, thus facilitating the adsorption process and making it more viable economically.

   The apparent significant differences of the adsorption properties of some microorganisms as a function of time are probably associated with a change in the morphological state of the cells. In the tests reported in Table 1, the cells were alive initially but were dead or inactive at the end of the tests [6,7]. In some cases, metal that initially was accumulated by the biomass could be later re-released into the

Table 1 Adsorption of Sr by Various Microorganisms During a 7-Day Period[a]

Microorganism	Distribution Coefficient[b]	
	2 Days	7 Days
Rhizopus arrhizus	1,470	85,533
Micrococcus luteus	24,943	12,010
Anabaena flos-aquae	5,554	10,304
Streptomyces viridochromogenes	4,147	3,949
Penicillium chrysogenum	7,181	3,625
Escherichia coli	967	1,961
Chlorella pyrenoidosa	734	1,915
Zooglea ramigera	493	1,858
Chlamydomonas reinhardtii	1,639	1,785
Candida sp.	764	1,139
Caulobacter fusiformis	619	1,055
Pseudomonas aeruginosa	6,321	1,003
Paecilomyces marquandii	1,808	656

[a]Tests were made in shake flasks at 30°C with washed live cells in deionized distilled water containing initially 10 ppm $SrCl_2$ buffered with 0.01 M phosphate buffer at pH 7 (results from Ref. 7).
[b]Distribution coefficient is defined as concentration of Sr in bioadsorbent on a dry weight basis (Sr concentration in solution).

solution. Such behavior was also observed by other researchers [10]. This behavior may be attributed to an initial, rapid Sr adsorption by a biomass constituent that is then slowly released from the cells. This may be the case if metal adsorption changed the solution pH causing structural changes in the biomass.

From such data, it is obvious that the amount of metal that can be accumulated by one type of microorganism may be vastly different from that which may be accumulated by another [7,11]. Thus, choosing the proper biological reagent for a particular adsorption system is not a trivial matter. In this regard, other researchers have shown that cells of nine species of the genus *Penicillium* had no statistically significant differences in their ability to accumulate metal ions [10]. Thus, metal accumulation properties may be more associated with a microbial species rather than with a genus.

Several researchers have demonstrated that bioadsorption behavior could be represented by typical adsorption isotherms [12,13]. Metal uptake by some microorganisms can be described by simple isotherms, such as those described by

Langmuir and Freudlich [12] while others must be described by more complex functions, such as the Brunauer-Emmett-Teller [BET] isotherm [13]. In evaluating the applicability of a given biomass for a particular processing situation, it is important to consider the behavior of the isotherm as a function of operating parameters, including the effect of solution concentration on maximum metal loading.

It has been shown that the rate at which metal ions are taken up by dispersed (free cells) microorganisms is quite rapid. For example, in 30 min $M.$ $luteus$ can remove greater than 98% of Sr at a concentration of 10 ppm [7]. Similarly, more than 60% of the equilibrium value of dissolved uranium and cobalt can be removed from solution by $S.$ $cerevisiae$ and $A.$ $nodosum$ during the first 15 min of contact [11], and over 99% of the Cd in solution can be adsorbed by $Aspergillus$ $oryzae$ with only a 5-min contact time [4].

In some cases, the rate of adsorption of metal ions into inactive or dead microbial biomass is both quite rapid and a function of the solution pH [14]. For example, the maximum adsorption rate for Cr^{6+} occurred at a pH of 2.5, where the rate is about twice that observed at a pH of 4. Conversely, the optimum rate of microbial adsorption of Cd^{2+} appears to be near pH 6 and it falls rapidly as the pH is lowered. In both cases, the maximum rate for bioadsorption is about the same at approximately 0.4 mg metal/g·min.

Plant Tissue

Although microbial biomass appears to be a relatively inexpensive medium for concentrating heavy metal ions, it is also possible to utilize the specialized structures contained in higher life forms to develop selective bioadsorbents that may have even higher capacities for some metal ions. In this regard, significant attention has recently been given to the use of plant tissue as the bioadsorbent.

Many plants are known to accumulate metals in various types of tissues. For example, root tissues of $Nicotiana$ $tobacum$ (tobacco) and $Lycopersicon$ $esculentum$ (tomato plants) are known to significantly concentrate many heavy metals [15–17]. Various other plant tissues have also been shown to concentrate metals, including $Brassica$ $oleracea$, $L.$ $esculentum$, Zea $mays$, and $Eichhornia$ $crassipes$ [5]. As in the case of microbial biomass and inorganic adsorbents, the plant tissues have typical adsorption isotherms (see Figure 1). The distribution coefficients for the bioadsorption of metals such as Sr by plant tissue can also be quite high, exceeding 500 for tomato roots and 150 for tobacco roots [17].

Live, growing plants [5,15,16], as well as inactive plant biomass [15], will adsorb metals. Even plant tissue cultures have been shown to adsorb Cd [5]. The system pH also appears to affect metal loading on plant biomass. In general, maximum loading occurs at higher pH while the adsorption distribution coefficient decreases at a lower pH. Thus, a decrease in the pH can be used to desorb or unload the accumulated metal.

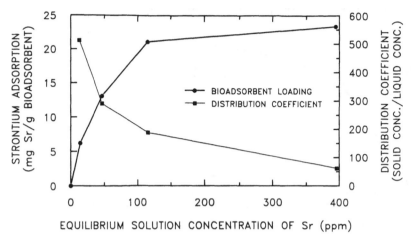

Figure 1   Isotherm and distribution coefficients for the bioadsorption of Sr by tomato root tissue at a pH of 7 and a temperature of 30°C. (From Ref. 17.)

Animal Tissue

One of the primary reasons that heavy metals are hazardous to higher animal life forms is because they tend to accumulate into particular animal tissue structures; for example, Sr tends to replace Ca in bones. These same types of biological materials can be used in a controlled system to selectively isolate dissolved metal pollutants or dilute resources. One such material is a modified bone gelatin that is composed of deionized bone gel (typically 15–25% in aqueous solution) which is crosslinked with up to 2 wt % propylene glycol alginate by the use of a dilute caustic solution [6,7]. This process results in a stable gel matrix that can potentially be used for a variety of processing concepts. It has been shown to be effective in removing trace amounts of Sr in aqueous waste streams even in the presence of large amounts of other low molecular weight divalent metal ions [6,7] such as Ca^{2+}.

Such material has also been shown to have a high adsorption capacity for dissolved Cu [12]. As in the case of other types of biomass, the Cu adsorption iostherms can be reasonably represented by the Langmuir isotherm. Also, a significant pH effect is observed in which essentially no adsorption occurs at a pH of 2 but bioadsorption capacity increases with increasing pH resulting in as much as 4 wt % Cu adsorbed at a pH of 5.5 (see Figure 2). This would essentially allow quantitative recovery of the metal ions by the modified bone gelatin by means of accumulation at a pH of 5.5 and subsequent recovery of the metal at high

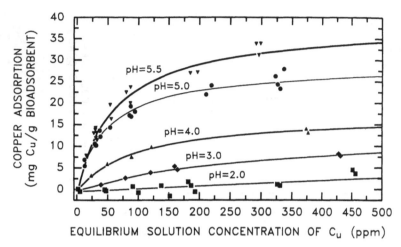

Figure 2 Bioadsorption isotherms for dissolved copper in modified bone gel. The solid lines represent the Langmuir approximation. (From Ref. 12.)

concentration by lowering the pH to 2. The bioadsorbent could then be reused to again adsorb metal at the higher pH.

The rate of adsorption in the modified bone gel is rapid when the stable gel matrix has a maximum dimension of about 1 mm. In this case, the time required to reach equilibrium loading is approximately 5 min [12].

ADSORPTION AND DESORPTION MECHANISMS

The mechanisms by which heavy metals are incorporated into biological systems are still not completely understood. Research on this subject has taken two primary approaches: (a) the study of bioadsorption mechanisms used by living microorganisms and tissues; and (b) investigation of biochemical processes associated with metal adsorption in nonliving or inactive biomass.

Adsorption by Living Organisms

Both active and passive mechanisms are employed by living cells to accumulate dissolved metals. Since high concentrations of heavy metals tend to adversely affect physiological processes, living cells that adsorb such materials must use mechanisms that ensure that only a limited amount of the metal actually reaches the interior of the cell where it is most toxic. Thus, many cellular mechanisms result in depositing the metals in the cell or membrane walls. For example, in the presence of a suitable phosphate donor, *Citrobacter* sp. is able to immobilize large quantities of Cd, Cu, Pb, and U by forming insoluble metal phosphates on the cell surface [18–20]. Similarly, sulfate-reducing bacteria may produce H_2S gas which will react with metal ions to form an insoluble precipitate [4]. If the cells are immobilized in a porous medium, then the precipitate containing the heavy-metal cations may actually form in the immobilization matrix.

Since many dissolved metals occur in a variety of oxidation states, biological systems could be used to convert the metal from a state that is difficult to remove from solution to one that is more amenable to solution detoxification. For example, bacterium such as *Enterobacter cloacae* will reduce chromate to trivalent chromium [21], which can then be removed using other bioadsorption technologies. To make this technology economically viable, however, the dynamics of this reaction must be enhanced significantly.

Phytochelatins, a class of peptides found in plants, are apparently effective in metal binding [5]. These peptides are in the range of 5–17 amino acids in length with a carboxyl terminal glycine that apparently binds metals by thiolate coordination. The phytochelatins are generated in living plant tissue that is exposed to metal ions, thus providing a means of enhancing the metal adsorption capacity of the biomass.

Adsorption by Nonliving or Inactive Biomass

Living cells are affected by metal toxicity, and they require the addition of necessary nutrients and the maintenance of specific operating conditions. On the other hand, nonliving or inactive biomass does not have these requirements, and in many cases it can be a very effective bioadsorbent for dissolved metals. The cell walls and internal materials within biomass offer particularly abundant sites for metal complexation or microprecipitation. Biological components have large amounts of polysaccharides, proteins, and lipids that are replete with metal-binding functional groups, including carboxylate, hydroxyl, sulfate, phosphate, and amino groups.

Since so many potential binding sites exist, it is difficult to identify clearly those that are specifically responsible for the primary binding of metals to cellular components. It has been shown, however, that in some algal biomass, ion exchange processes are the predominant mechanisms by which this biomass sequestered cobalt from aqueous solutions [22,23]. It was speculated that the biomass and the ions probably form a crosslinked network, similar to that formed in alginate gels in the presence of suitable metal cations. As for the case of more conventional adsorbers in which ion exchange mechanisms are known to occur, competition for the binding sites by different metal ions also occurs in biomass [10,24].

In the biomass formed from *Rhizopus arrhizus*, it was shown that the amount of metal that could be bound was linearly related to the ionic radius of the metals, with larger ions being adsorbed more strongly than smaller ions [24]. This behavior was explained by complexation between the metal and the functional sites in the biomass.

It has been shown by several researchers that for some types of microbial biomass, most of the metal adsorption occurs in the cell or membrane wall. This included biomass from fungi such as *Ganoderma lucidum*, which adsorbs Cu [25]; bacteria such as *M. luteus*, for the adsorption of Sr [9]; and *Saccharomyces cerevisiae* for the adsorption of U [26]. In the case of *M. luteus*, like that of other gram-positive bacteria, the high peptidoglycan content, of which glutamic acid is a major component, may be responsible for Sr binding.

Apparently, extracellular capsular materials with carboxylic residues show higher uranium sorption capacity than do whole cells [27]. In such extracellular materials, more potential binding sites exist per unit of material than in cellular biomass. Similarly, it was determined that the uptake capacities of melanin were greater than intact fungal biomass [13].

Accumulation of cobalt metal ions by *Ascophyllum nodosum* has been shown to be associated with bonds $-C=C-$ bond stretching and/or a shift in amino groups and bonds $-CH-$ and/or bonds $-COOH$ stretching. In either case, ion exchange must play an important role in the initial adsorption and in the desorption [28].

It has been postulated that in the bioadsorption of Cu by crosslinked bone gelatin, counterion condensation may be involved [35]. Interestingly, a significant size change is associated with this adsorption mechanism.

As previously mentioned, metal-chelating thiolate groups that are common in proteins have been shown to bind some metals. This has resulted in the design and use of a synthetic cadmium-binding adsorbent in which similar active sites were attached to a commercially available ion exchange resin that had a very high capacity for the metal [29].

Desorption Mechanisms

Since bioadsorption of metals appears to be an equilibrium process primarily controlled by ion exchange or chemical complexation, it would be expected that a modification to the chemical environment could be used to induce desorption of the metals. That is, a change in the chemistry can be used to significantly decrease the affinity of the metal for the bioadsorbent. It has been demonstrated that displacement by competing ions and reversal of the complexation reaction can be effective for this purpose. As previously shown, the desorption of Cu from modified bone gel can be achieved by simply decreasing the pH from 5.5 to 2 (Figure 2) [12]. It has also been shown that bioadsorption associated with ion exchange sites can be reversed by increasing the concentration of a competing ion. An example of the latter approach is the use of high concentrations of $CaCl_2$ to desorb Co from *A. nodosum* [28]. Such desorption processes can be used to concentrate and recover the metal values and reuse the bioadsorbent in a recycle mode.

Since the biomass is relatively inexpensive, it might be preferable to simply dehydrate the loaded bioadsorbent and use it to isolate the metals for disposal. The volume can also be further decreased by incineration, since this is an organic rather than an inorganic adsorbent, and the resulting ash will have even higher concentrations of the metal.

BIOPROCESSING SYSTEMS

The bioadsorbent can be used by dispersing it in free suspension in stirred-tank contactors with subsequent filtration for recovery of the metal [3]. But it is probably even more effective to immobilize the biomass into cohesive particles that can be used in columnar-type contactors with the advantage of multistage operation [4,6,7].

Immobilization of the Bioadsorbent

Bioadsorbent material can be immobilized as particulates by chemical crosslinking, as in the case of modified bone gelatin or unprocessed biomass [2,18], or by incorporation onto or into a particle or gel matrix (Figure 3) [4,30]. Microbial

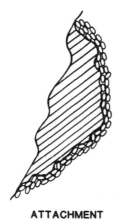 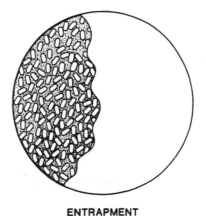

ATTACHMENT	**ENTRAPMENT**
(TYPICALLY A BIOLOGICAL FILM ON A SOLID SURFACE)	(TYPICALLY MICROORGANISMS ENCAPSULATED IN A GEL MATRIX)

Figure 3 Two primary types of the immobilization of biomass. (From Ref. 30.)

films can be induced to form on the external surfaces of some types of particulates, thus producing fixed biomass that may even be stable in the high-shear fields of fixed- or fluidized-bed contactors [4,31].

The entrapment of biomass into gel matrices allows very large quantities of the biomass to be included in columnar contactors. For example, biomass can be up to 60% of the total volume of "bioadsorbent" beads made by incorporating microbial mass into gel media such as carrageenan, alginate, or modified bone gel [32]. When the inactive biomass itself is chemically stabilized into particulates, the biomass content of a columnar reactor may be greater than 25% of the total contactor volume [6,7]. These latter two techniques appear to provide the maximum amount of bioadsorbent contact for most applications. As described earlier, diffusional limitations do not appear to be significant for this immobilized material as long as the diffusional path is <1 mm.

Contacting Systems

Desirable characteristics of an ideal bioadsorbent contacting system will likely include continuous operation, high bioadsorbent concentrations, and effective contact between the bioadsorbent and the interacting constituents. Columnar

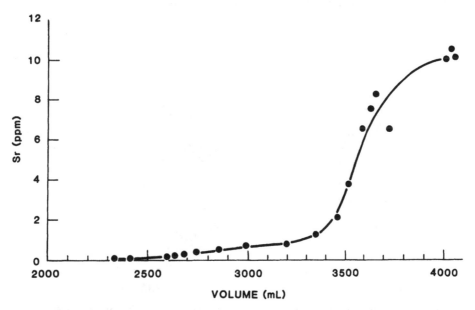

Figure 4   Effluent concentration histogram of Sr from a nominal 50-ml fixed-bed column of crosslinked bone gelatin containing 20 vol % of *M. luteus* biomass that is contacted by an aqueous stream containing 10 ppm $SrCl_2$. (From Ref. 7.)

configurations utilizing a fixed bed or fluidized bed of bioadsorbent particles appear to be the most attractive approach. Such systems are not yet extensively used commercially, but they are beginning to be investigated for several applications.

When relatively long retention times with high bioadsorbent capacity are required, a fixed-bed configuration with relatively large particles (0.5–4 mm diameter) will probably be the choice. Such systems have been shown to be effective for the removal of U from dilute solutions [33] and for the uptake of Sr [6,7]. In the latter case, loading and ultimate breakthrough of the dissolved metal occurs in a fixed bed of particles of immobilized bone gel containing 20 vol % of *M. luteus* (see Figure 4). The bioadsorbent in the fixed bed can then be regenerated by a significant increase in KCl salt concentration (see Figure 5). The resulting recovered Sr can be concentrated by more than a factor of 10 by this technique.

Other columnar contacting systems have also been studied. A unique counter-current approach was considered for the removal of trace quantities of U from

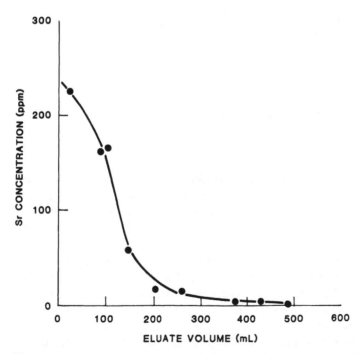

Figure 5 Desorption of Sr from a nominal 50-ml fixed-bed column of crosslinked bone gelatin containing 20 vol % of *M. luteus* biomass that is contacted with 0.3 M KCl solution. (From Ref. 7.)

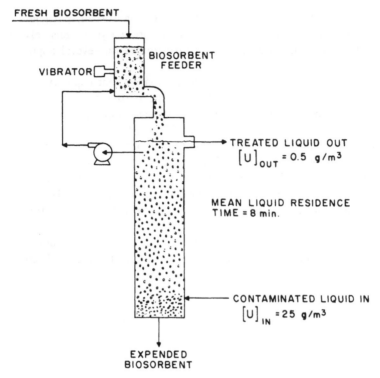

FRESH BIOSORBENT

VIBRATOR

BIOSORBENT
FEEDER

TREATED LIQUID OUT
$[U]_{OUT} = 0.5 \ g/m^3$

MEAN LIQUID RESIDENCE
TIME = 8 min.

CONTAMINATED LIQUID IN
$[U]_{IN} = 25 \ g/m^3$

EXPENDED
BIOSORBENT

Figure 6   Columnar countercurrent contactor used for continuous removal of uranium by cascading particles of anthracite coal that have an attached film of microbial biomass. (From Ref. 34.)

solution (see Figure 6) [1,34]. In this concept, there is a continuous upflow of the aqueous stream that is contacted countercurrently by cascading particulates of anthracite coal that include a film of microbial biomass. It was shown that 98% of dissolved U at a concentration of 25 ppm could be removed during a contact time of only 8 min.

Another unique columnar system has been proposed for the use of crosslinked bone gel beads [35]. With such particulates, a significant size reduction occurs as metal ions are accumulated. It has been speculated that these changes could be due to ionic interactions with a chelating polyelectrolyte structure. This structure would then be responsible for not only the size changes associated with the adsorption of metal ions but also for the adsorption phenomena itself. A fluidized-bed contactor could be used to facilitate a self-imposed separation of the smaller loaded beads from the rest of the bed, since they will migrate to the top of the bed for easy removal. Then, the loaded beads could easily be removed and regenerated.

FUTURE DEVELOPMENTS

The use of bioadsorbents for the removal or recovery of dissolved metals has been demonstrated for a variety of applications on a laboratory scale or with small pilot plants. Commercial applications are just beginning, but they will likely multiply in a very few years. In order to bring this about, efficient forms of bioadsorbents must be available and used in contacting systems that are continuous and have high capacity. Thus, future research will undoubtedly result in a variety of immobilized bioadsorbent particulates or beads that are highly concentrated with appreciable capacity. Although microbial biomass has been the primary type of bioadsorbent tested to date, additional concepts utilizing plant and animal materials will allow many different future concepts.

The most successful contacting systems for use of the immobilized bioadsorbents will be continuous columnar systems. Most future industrial bioadsorption systems will probably be based on fixed-bed contactors; however, other specialized columnar systems utilizing fluidized beds or other concepts will also find a place.

The majority of future applications will probably be for the removal of metal pollutants in aqueous effluents. But the recovery and concentration of valuable metals from dilute aqueous processing streams could also result in important applications.

ACKNOWLEDGMENTS

We gratefully acknowledge the support of the Division of Engineering and Geosciences, Office of Basic Energy Sciences, U.S. Department of Energy for much of the research and for the preparation of this chapter.

REFERENCES

1. S. E. Shumate and G. W. Strandberg, *Comprehensive Biotechnology*, Vol. 4 (C. L. Cooney and A. E. Humphrey, eds.), Pergamon Press, New York, 1985, p. 235.
2. J. A. Brierley, G. M. Goyak, and C. L. Brierley, *Fundamental and Applied Biohydrometallurgy* (R. W. Lawrence, R. M. R. Bronien, and H. G. Ebner, eds.), Elsevier, New York, 1986, p. 291.
3. B. Volesky, *Trends Biotechnol. 5*, 96 (1987).
4. L. E. Macaskie and A. C. R. Dean, *Advances in Biotechnological Processes*, Vol. 12 (A. Mizrani, ed.), A. R. Liss, New York, 1989, p. 159.
5. E. Grill, E.-L. Winnacker, and M. H. Zenk, *Proc. Natl. Acad. Sci. 84*, 439 (1987).

6.   J. Watson, C. D. Scott, and B. D. Faison, *Emerging Technologies in Hazardous Waste Management*, ACS Symposium Series No. 422 (D. W. Tedder and F. G. Pohland, eds.), American Chemical Society, Washington, 1990, p. 173.
7.   J. S. Watson, C. D. Scott, and B. D. Faison, *Appl. Biochem. Biotechnol. 20/21*, 699 (1989).
8.   M. Tsezos, M. H. I. Baird, and L. W. Shemilt, *Chem. Eng. J. 32*, B29 (1986).
9.   B. D. Faison, C. A. Cancel, S. N. Lewis, and H. I. Adler, *Appl. Environ. Microbiol. 56*, 3649 (1990).
10. L. Pighi, T. Pumpel, and F. Schinner, *Biotechnol. Lett. 11*, 275 (1989).
11. N. Kuyucak and B. Volesky, *Biotechnol. Lett. 10*, 137 (1988).
12. J. N. Petersen, B. H. Davison, C. D. Scott, and S. L. Blankinship, *Biotechnol. Techniques 4*, 435 (1990).
13. F. M. Gadd and L. de Rome, *Appl. Microbiol. Biotechnol. 29*, 610 (1988).
14. Y. Sag and T. Kutsal, *Biotechnol. Lett. 11*, 145 (1989).
15. H. Lue-Kim and W. E. Rauser, *Plant Physiol. 81*, 896 (1986).
16. G. S. Wagner and R. Yeargah, *Plant Physiol. 82*, 274 (1986).
17. C. D. Scott, *Biotechnol. Bioeng. 39* (1992).
18. L. E. Macaskie and A. C. R. Dean, *Biotechnol. Lett. 7*, 457 (1985).
19. L. J. Michel, L. E. Macaskie, and A. C. R. Dean, *Biotechnol. Bioeng. 28*, 1358 (1986).
20. L. E. Macaskie, J. M. Wates, and A. C. R. Dean, *Biotechnol. Bioeng. 30*, 66 (1987).
21. K. Komori, A. Rivas, K. Toda, and H. Ohtake, *Biotechnol. Bioeng. 35*, 951 (1990).
22. N. Kuyucak and B. Volesky, *Biotechnol. Bioeng. 33*, 809 (1989).
23. N. Kuyucak and B. Volesky, *Biotechnol. Bioeng. 33*, 823 (1989).
24. J. M. Tobin, D. G. Gooper, and R. J. Neufeld, *Appl. Environ. Microbiol. 47*, 821 (1984).
25. T. R. Muraleedharan and C. Venkobachar, *Biotechnol. Bioeng. 35*, 320 (1990).
26. G. W. Strandburg, S. E. Shumate, and J. R. Parrott, *Appl. Environ. Microbiol. 41*, 237 (1981).
27. J. M. Scharer and J. J. Byerley, *Hydrometallurgy 21*, 319 (1989).
28. N. Kuyucak and B. Volesky, *Biotechnol. Bioeng. 33*, 815 (1989).
29. J. Yin and H. W. Blanch, *Biotechnol. Bioeng. 34*, 180 (1989).
30. C. D. Scott, *Enzyme Microb. Technol. 9*, 65 (1987).
31. C. D. Scott, C. W. Hancher, and E. J. Arcuri, in *Advances in Biotechnol*, vol. 1 (M. Moo-Young, ed.), 1981, p. 651.
32. C. D. Scott, C. A. Woodward, and J. E. Thompson, *Enzyme Microb. Technol. 11*, 258 (1989).
33. M. Tsezos, R. G. L. McCready, and J. P. Bell, *Biotechnol. Bioeng. 34*, 10 (1989).
34. S. E. Shumate, G. W. Strandberg, D. A. McWhirter, J. R. Parrott, G. M. Bogacki, and B. R. Locke, *Biotechnol. Bioeng. Symp. 10*, 27 (1980).
35. J. N. Petersen, B. H. Davison, C. D. Scott, and S. L. Blankinship, *Biotechnol. Bioeng. 38*, 923 (1991).

5

Membrane Separations in the Recovery of Biofuels and Biochemicals: An Update Review

Stephen A. Leeper*   *Idaho National Engineering Laboratory, Idaho Falls, Idaho*

Several factors are creating increased interest in the production of fuels and chemicals via bioconversion. Bioconversion provides a route to organic chemicals and fuels that does not depend on petroleum feedstocks. Increasing petroleum prices have occurred in the past and are anticipated in the future; the supplies of petroleum are decreasing and disruptions to supplies can occur anytime. Bioconversion provides a new potential market for grain products. The environmental effects of bioconversion are relatively benign by comparison to synthetic (petroleum-based) chemical conversion (Busche, 1989; Young, 1989). The production of fuels and chemicals from biomass via bioconversion could decrease the net production of greenhouse gases. In addition, bioconversion provides an alternative method for disposing of wastes; many industrial wastes are suitable feedstocks for bioconversion. Recent developments in molecular biology and in separations technology have also increased the potential of bioconversion for the production of fuels and chemicals. The pertinent issues on this topic have been described and evaluated by Leeper and Andrews (1991) and Leeper et al. (1991). However, product recovery costs remain a major barrier to increased use of bioconversion for the production of fuels and chemicals. Advances in separation technology are vital to the successful development of a bioconversion-based fuels and chemicals industry. Most organic chemical products of bioconversion are produced in low concentration in water. Recovery of these products is energy-

---

Current affiliation: Clinton Laboratories, Eli Lilly and Company, Clinton, Indiana.

intensive and expensive. The integration of membrane technology separations into product recovery processing may reduce the energy requirements and costs of product recovery in bioconversion.

In this vein, the status of membrane technology for the recovery of fuels and organic chemicals produced via bioconversion is reviewed. This review is an update of a previous review on the subject of membrane separations in the production of alcohol fuels (Leeper, 1986). In this chapter, data and literature published since the previous review are compiled and discussed.

Products that are discussed in this review include ethanol, acetone, butanol, isopropanol, and other organic fuels/chemicals produced by fermentation. Separation technologies that are discussed include pervaporation (PV), vapor permeation (VPe), reverse osmosis (RO), membrane extraction (perstraction), and electrodialysis (ED). Ultrafilatration (UF) is not discussed in this chapter. The literature on UF in biotechnology is expansive. In addition, the primary uses of UF in biotechnology involve concentration of cells, clarification of fermentation broth, and retention of cells in membrane bioreactors. UF applications in biotechnology are discussed by Strathmann (1985), Klinkowski (1983), Michaels and Matson (1985), Cheryan (1986), and Eykamp and Steen (1987). This chapter is limited to uses of membrane separation operations that directly increase the purity of fuels and organic chemicals produced by fermentation.

MEMBRANE UNIT OPERATIONS

In recent years, the use of membrane technology in industrial separations has been increasing, especially in water treatment and food processing. The industrial uses of membrane technology have been reviewed (Leeper et al., 1984). A review of applications in the food industry is also available (Mohr et al., 1988). A comprehensive evaluation of research needs has been prepared (DOE, 1990). Hwang and Kammermeyer (1984) and Rautenbach and Albrecht (1989) have provided excellent reference texts on membrane theory and unit operations. The membrane unit operations discussed in this chapter are briefly described and the terms used to measure membrane system performance are briefly defined.

Reverse Osmosis (RO)

In RO, a purified liquid is generally separated from the feed solution, which contains solutes (usually salts, sugars, proteins, etc.) or other liquids (e.g., ethanol and water). The product can be the purified fluid that flows through (permeates) the membrane (e.g., desalting of water). Or the concentrated fluid stream that is retained by the membrane (retentate) can also be the product (e.g., ethanol concentration). The separation is driven by a pressure difference across the membrane. Operating pressures of 2.0–7.0 MPa are typical in RO. RO is described in greater detail by Sourirajan and Matsuura (1985), Petersen (1986), and Eykamp and Steen (1987).

Pervaporation (PV)

In PV, the feedstream is a liquid mixture and a partial vapor pressure difference is maintained across the membrane. The permeate side of the membrane is kept at vacuum pressures or is flushed with an inert (sweep) gas stream (e.g., nitrogen). Some components preferentially permeate across the membrane and vaporize on the low-pressure (permeate) side. The permeate side vapor is condensed and recovered. The less permeable components are concentrated in the retentate stream. Pervaporation is reviewed by Ishida and Nakagawa (1985), Rautenbach and Albrecht (1985a, 1985b, 1987) and Bruschke (1988, 1990). A brief review of transport in PV is provided by Blume et al. (1990). A systematic definition of the terms used in PV, with a brief description of transport models, is provided by Boddeker (1990).

Vapor Permeation (VPe)

Vapor permeation is similar to pervaporation. In VPe, the feedstream is a mixture of vapors. The permeate side of the membrane is maintained at vacuum pressures or is flushed with an inert (sweep) gas stream. The partial vapor pressure difference across the membrane provides a driving force for selective vapor permeation. Vapor permeation is reviewed by Meares (1988).

Membrane Extraction (Perstraction)

Membrane extraction is similar to liquid–liquid extraction, except that a membrane is used as a barrier between the feedstream and the solvent stream. The feedstream contains a solute or liquid to be recovered. A solvent stream flows on the permeate side of the membrane. The desired solute or liquid in the feedstream selectively permeates across the membrane and into the solvent stream. The chemical potential difference across the membrane creates the driving force. Perstraction is described in greater detail by Cabasso et al. (1974) and Matsumura and Markl (1984).

Electrodialysis (ED)

In ED, ions are separated from solvents and from solutions of neutral solutes by ion exchange membranes and an applied electric field (the driving force). Ion exchange membranes reject ions on the basis of charge; for instance, an anion exchange membrane contains positively charged molecules and rejects cations, but does not reject anions. In an ED cell, anion and cation exchange membranes are arranged in an alternating sequence. When the electric field is applied, for instance, cations migrate toward the negative pole until an anion exchange membrane is encountered. In this way, cations and anions are concentrated in specific channels and are separated. ED was recently reviewed by Klein et al. (1987).

Membrane Performance (Terminology)

The terminology used in membrane technology is not universal. Care should be taken when reading the literature. A standardized approach to membrane technology terms was proposed by Hallstrom and von Sengbusch (1986).

The terms as used in this chapter to quantify membrane performance are defined below. *Permeability* and *flux* describe separation rate. *Separation factor* and *rejection* describe the degree of separation.

Permeability is a basic property of membrane materials or membrane systems. Permeability is usually presented as a property that is independent of membrane thickness and driving force. Use of this term is limited in this chapter. For a thorough discussion of permeability, see Hwang and Kammermeyer (1984), Sourirajan and Matsuura (1985), or Rogers and Sfirakis (1986). For a brief discussion, see Mohr et al. (1988).

For the design of membrane operations, *flux* is generally more useful than permeability. Flux is generally considered to be the bulk fluid transport rate, but the flux of specific components can also be measured. Flux is generally expressed in units of quantity of fluid permeating per unit membrane area per unit time. In this chapter, flux is reported as $kg/m^2 \cdot hr$ (abbreviated KMH).

Separation factor (α_{AB}) is the term used to measure separation in pervaporation and vapor permeation. This term (as used in this chapter) is defined as:

$$\alpha_{AB} = \frac{Y_A/Y_B}{X_A/X_B}$$

where

   Y = average permeate side concentration of component A or B
   X = average feed side concentration of component A or B [or (sometimes) concentration in the feed]

Many authors call the above term *selectivity*. However, selectivity is more properly defined as the ratio of component permeabilities. The values of these two terms approach each other for systems in which the permeate side pressure approaches zero and the feed side concentrations remain essentially constant (i.e., when the stage cut approaches zero).

Rejection is the term used to measure separation in reverse osmosis. This term is defined as:

$$R_i = \frac{X_i - Y_i}{X_i} \times 100\%$$

where

X_i = average feed side concentration of component i [or (sometimes) feed concentration].

Y_i = average permeate side concentration of component i

Rejection is also called retention or, if expressed as a fractional value, retention coefficient.

These measures of membrane performance are dependent on the system in which they were measured and the measurement conditions. They can be differential or integral values (Leeper, 1986). The original paper should always be consulted for membranes of interest.

NOTES ON THE TABLES

A few comments will aid understanding of the tables. In the tables, references are cited by first author only (unless this shorthand method will cause confusion). References cited in the tables are not generally repeated in the text.

The symbols used in the tables are defined as follows:

α = separation factor

A = acetone or component A

B = butanol or component B

E = ethanol

J = flux

NR = not reported

P = isopropanol

R = rejection

T = temperature, or total (if subscript)

W = water

X = weight percent of indicated component in feed

Y = weight percent of indicated component in permeate

Additional terms used in the tables are defined in the "Comments" for that membrane. The various conditions and properties of the membranes are reported as ranges.

For the tables providing pervaporation data, *water-selective membrane* means that water is the component that preferentially permeates (etc.). Presentation of PV data in McCabe–Thiel–type (y vs. x) diagrams is becoming common; these diagrams do not provide equilibrium data, but they are useful for visualizing the effectiveness of specific membranes and for determining the concentration ranges in which use of a specific membrane is likely to be advantageous. For data presented in this format, the following terms are used in the tables to describe separation factor: low ($\alpha \sim 1-5$); moderate ($\alpha \sim 5-25$), high ($\alpha \sim 25-100$), very high ($\alpha > 100$).

Additional pertinent information is provided under "Comments."

ETHANOL RECOVERY

The production of ethanol by fermentation of grain has increased dramatically since the mid-1970s, when essentially all ethanol was produced synthetically. In 1988, U.S. ethanol production by fermentation was 3.1 million m³ (828 million gallons) (Hagwood, 1989). For comparison, U.S. synthetic ethanol production was 0.33 million m³ in 1988 (Anonymous, 1989). Ethanol will probably also be made from lignocellulosic feedstocks in the future. The ethanol industry is growing; new and advantageous technology developments will be incorporated as they occur. Advanced separations for the recovery of ethanol will facilitate the development of new projects for production of ethanol from grain and lignocellulose.

Ethanol production by fermentation was briefly reviewed in Leeper (1986) and is not repeated here. The traditional process for production and recovery of beverage ethanol is described by Maisch et al. (1979). Recent reviews on ethanol production from lignocellulose and xylose include Jeffries (1985), Maiorella (1985), Schneider et al. (1986), Parisi (1989), and Lynd (1989, 1990). Developments in ethanol recovery have been evaluated by Collura and Luyben (1988) and reviewed by Maiorella (1985) and Serra et al. (1987). Various separations are being studied for the recovery of ethanol, including advanced distillation (Collura and Luyben, 1988), vapor recompression distillation (Tegtmeier, 1985), extractive distillation (Lee and Pahl, 1985a), liquid extraction (Kollerup and Daugulis, 1985b; Crabbe et al., 1986; Egan et al., 1988), liquid extraction using gasoline as the solvent (Lee and Pahl, 1985b; Leeper and Wankat, 1982), extractive fermentation (Kollerup and Daugulis, 1985a, 1985b; Fournier, 1986; Honda et al., 1987), extraction/phase separation (Mehta and Fraser, 1985), CO_2 stripping (Pham et al., 1989), and sorption (Sinegra and Carta, 1987). Of course, membrane separations are also being studied; citations are given in the following sections.

The energy requirements (as presented in the literature) for several ethanol purification processes are provided in Table 1. Conventional distillation processes for purification of ethanol from grain-based fermentation broths (containing 7–10 wt % ethanol) consume 6.7–8.2 GJ/m³ of ethanol (anhydrous basis). Other nonmembrane processes include supercritical extraction plus azeotropic distillation (>4.5 GJ/m³), conventional distillation plus dehydration by adsorption with cornmeal (3.3 GJ/m³), extraction plus distillation and azeotropic distillation (4.9 GJ/m³), and conventional distillation plus extraction with gasoline to directly produce gasoline/ethanol blends (5.5 GJ/m³). Vapor recompression distillation, an advanced distillation operation, consumes about 2.5 GJ/m³. References for these data are provided in Table 1.

Pervaporation

Pervaporation is being extensively studied for dehydration of ethanol and for the preconcentration of dilute ethanol. PV is attractive for several reasons (Frennesson et al., 1986; Sander and Soukup, 1988a; Asada, 1988; Fleming, 1989b). Azeotropic mixtures can be separated relatively easily using PV. PV is a low-

Table 1 Energy Requirements of Ethanol Recovery

Process	Stream concentration (wt %)		Energy consumption		Ref.
	(Feed)	(Final)	(GJ/m³)	(Btu/gal)	
Reverse osmosis	4.0	10.0	0.64	2,300	Bitter (1988)
Conventional distillation	10.0	95.0	4.7	17,000	Serra (1987)
Azeotropic distillation	95.0	99.9	2.0	7,200	Serra (1987)
Total	4.0	99.9	7.3	26,500	
Reverse osmosis	2.5	8.0	0.30	1,100	
Conventional distillation	8.0	92.4	6.64	23,800	
Azeotropic distillation	92.4	99.5	1.95	7,000	Leeper and Tsao
Total	2.5	99.5	8.89	31,900	(1987)
Conventional distillation	2.0	92.4	22.32	80,100	
Azeotropic distillation	92.4	99.5	1.95	7,000	Leeper and Tsao
Total	2.0	99.5	24.27	87,100	(1987)
Conventional distillation	7.0	92.4	6.2	22,200	Collura (1988)
Azeotropic distillation	92.4	99.5	1.95	7,000	Leeper and Tsao
Total	7.0	99.5	8.2	29,200	(1987)
Conventional distillation	10.0	95.0	4.74	17,000	Serra (1987)
Azeotropic distillation	95.0	99.9	2.0	7,200	Serra (1987)
Total	10.0	99.9	6.74	24,200	
Vapor recompression distillation	10.0	95.0	1.70	6,100	Serra (1987)
Azeotropic dist (ether)	95.0	99.9	0.84	3,000	
Total	10.0	99.9	2.54	9,100	
Supercritical extraction w/CO₂	10.0	91.0	2.51	9,000	
Azeotropic distillation	92.5	99.9	>2.01	>7,200	
Total	10.0	99.9	>4.5	>16,200	Serra (1987)
Conventional distillation	8.0	80.0	—	—	
Pervaporation	80.0	99.5	—	—	
Total	8.0	99.5	3.68	13,200	Serra (1987)
Conventional distillation	10.0	87.5	—	—	
Water adsorption (cornmeal)	87.5	99.5	—	—	
Total	10.0	99.5	3.34	12,000	Serra (1987)
Two step extraction	10.0	26.0	—	—	
Conventional distillation	26.0	98.0	—	—	
Total	10.0	98.0	4.90	17,600	Serra (1987)
Conventional distillation	10.0	90.0			
Extraction with gasoline to produce gasohol	90.0	10.0	5.52	19,800	Leeper and Wankat (1982)

energy operation compared to distillation, since only a fraction of the feed (the permeate) is vaporized. PV equipment is compact, easily automated, and easy to operate. Startup and shutdown are relatively simple. PV is flexible; the same equipment can perform different separations and can handle changes in feed composition. Construction can be rapid and installation costs are low since PV units are prefabricated. At small scale, PV can have lower capital and operating costs than azeotropic distillation. The use of PV for ethanol purification was reviewed previously (Leeper, 1986; Frennesson et al., 1986; Bruschke, 1988).

Pervaporation: Ethanol Dehydration (Water-Selective Membranes)

PV is especially attractive for the dehydration of organic chemicals. Multicomponent mixtures can be dehydrated because with appropriate membranes water is more permeable than organics. The product is free of a third component, since a third component is not needed to effect separation. Recovery of an entrainer is not necessary, as in azeotropic distillation. PV is being used for ethanol dehydration in pilot and commercial installations in Europe (Bruschke, 1988), Asia (Fleming, 1989b), and Japan (Asada, 1988). PV processes, energy requirements, and economics are discussed after ethanol/water pervaporation data are presented.

The major commercial pervaporation membrane is the GFT polyvinyl alcohol membrane. This membrane consists of three polymer layers. The separation layer (i.e., the layer that controls flux and selectivity) consists of polyvinyl alcohol (PVA) and has a thickness of about 0.5 μm. The support layer is provided by polyacrylonitrile with an asymmetric, highly porous structure about 100 μm in thickness. These two layers exist on a nonwoven polyester fabric which provides additional mechanical strength to the membrane (~200 μm). Surprisingly, for this membrane, it has been reported that flux is independent of membrane thickness (Sander and Soukup, 1988a).

Ethanol/water pervaporation data for PVA membranes, which are water-selective, are presented in Table 2. Detailed and complete ethanol/water pervaporation data are available for PVA membranes. The separation factors exhibited by PVA are very high at high ethanol concentrations, which makes PVA well suited for use in ethanol dehydration. For dehydration applications (in which the ethanol concentration is greater than 90 wt %), PV flux with PVA membranes is greater than $0.1 \text{ kg/m}^2 \cdot \text{hr}$ at temperatures greater than 100°C. Mass transport through PVA membranes has been modeled by Hauser et al. (1989), using an approach based on nonideal solubilities of the components; this model shows much greater agreement with experimental data than do models based on the approximation of ideal (additive) component solubilities.

Cellulose acetate (CA) membranes have also been extensively studied for PV. PV data for CA membranes are presented in Table 3. CA membranes tend to be characterized by low to moderate separation factors and relatively low flux. However, carboxymethylcellulose (CMC) membranes exhibit very high separation factors, but very low flux. Notable exceptions may include the CMC/ polyacrylic acid ion exchange membrane and the cellulose nitrate/polymethyl acrylate membrane.

Ethanol/water pervaporation data for polysulfone membranes are presented in Table 4. These membranes tend to exhibit low separation factors and low flux.

Data for charged polyacrylic acid (PAA) and acrylate membranes are presented in Tables 5 and 6. The performance of these membranes is highly dependent on the counterion used in the membrane. Membranes with sodium cations exhibit both high flux and moderate to high separation factors. For instance, the PAA/polypropylene graft polymerized copolymer (GPC) membrane has low separation factors and good flux, whereas the PAA/polypropylene GPC membrane (with Na$^+$ counterions) exhibits moderate to high separation factors and higher flux (Hirotsu, 1987a). The modified PAA membranes, especially the PAA/PCA-107 membrane, look promising.

Fluorinated materials are also being studied. Data for polyvinyl fluoride (PVF), polyvinylidene fluoride (PVDF), and other fluorinated membranes are presented in Tables 7–9. The copolymers with PAA appear to exhibit the highest flux and separation factors. However, these membranes need to be studied at higher ethanol concentrations before conclusions can be drawn about their potential utility. Many of these membranes have been reported to be unstable. The PVF/4-vinylpyrrolidone (bètaine) membrane (Table 7) was stable and exhibited good performance. A radiation-cured PVDF membrane (Table 8) also exhibits impressive performance. Nafion membranes (Table 9) exhibit low separation factors at high ethanol concentration and do not exhibit especially high flux. The polyhydroxymethylene-co-fluoroolefin membrane exhibits very high separation factors but appears to have low flux.

Polyacrylonitrile (PAN) membranes are being studied (Table 10). Again, some of the more interesting PAN membranes are copolymers with PAA, with sodium and potassium counterions. Radiation-cured PAN membranes exhibit very high flux and separation factors.

Ethanol/water PV data for miscellaneous membranes are presented in Table 11. The styrene/butadiene/polyvinyl chloride membranes may exhibit interesting performance. Cation exchange membranes exhibit moderate separation factors and high flux; these membranes may prove to be useful. The transport mechanisms of PV with ion exchange membranes are discussed by Cabasso (1985a, 1985b, 1986). At high ethanol concentrations, the sulfate anion exchange membrane has performance that may be equivalent to PVA membranes.

The general schematic for ethanol dehydration by PV is shown in Figure 1. The hybrid distillation/PV concept is discussed by Gooding and Bahouth (1985). The use of PV in hybrid distillation/PV processes can have lower capital and operation costs than the use of either operation alone. In addition, PV is generally a staged process to allow feed/retentate stream reheating due to energy loss by permeate vaporization. The plate-and-frame Lurgi pervaporator, which has a

Table 2 Ethanol/Water Pervaporation Data: Polyvinyl Alcohol Membranes (Water-Selective)

Membrane material	Type of data	Ethanol concentration in feed (wt %)	Temperature (°C)
Polyvinyl alcohol (GFT)	J_W vs. X_E	92–100	90–100
	Y_W vs. X_E	0–100	100
	Y_E vs. X_E	90–100	100
Polyvinyl alcohol (GFT)	J_T vs. X_E, T Y_E vs. X_E, T α vs. X_E, T J_T, α vs. PP	0–100	60
Polyvinyl alcohol	J_T vs. X_E α vs. X_E	5–95	70
Polyvinyl alcohol	Y_E vs. X_E	0–100	60
Polyvinyl alcohol	Y_E vs. X_E	0–100	
	J_T vs. X_E	0–100	
	J_T vs. X_E	70–100	60
	J_T vs. X_E	90–100	
Polyvinyl alcohol (homogeneous)	J_T vs. X_W α vs. X_W	70–100	25
Polyvinyl alcohol (with maleic acid crosslinking)	J_T vs. X_E α vs. X_E	80–95 80–95	70
	J_T, α vs. T	80	40–70
Polyvinyl alcohol (with amic acid crosslinking)	J_T vs. X_E α vs. X_E J_T vs. T α vs. T	30–90 10–90	NR
Crosslinked polyvinyl alcohol	J_T vs. X_W α vs. X_W	50–90	40
Surface-modified polyvinyl alcohol	J_T vs. X_W α vs. X_W	50–90	40

Permeate pressure (kPa)	Separation factor (α_{WE})	Flux (kg/m²·hr)	Comments	Ref.
NR	High	~0–0.9	GFT is manufacturer of PVA pervaporation membrane.	Sander (1988a)
	Comments		For X_E < 25 wt % ethanol, PVA is ethanol-selective at low α.	
	High		For X_E > 25 wt % ethanol, PVA is water-selective at low to high α.	
0.2	High	0–4	PP is permeate pressure. J_T and α highest at lowest PP.	Wesslein (1988)
	1–300		Water permeates for X_E > 40 wt %.	Wesslein (1990)
2–200				
	2–16	~0–1.0		Changlou (1988)
2.0	Comments	NR	α high for X_E > 75 wt %. Experimental and predicted data.	Hauser (1988)
			Model approximates experimental data.	Hauser (1989)
0.01	Comments		α high for X_E > 75 wt %.	Hauser (1987)
		0–2.4	Flux is highest at high X_W.	
		0–1.2	Model approximates experimental	
		0–0.1	data for X_E > 70 wt %.	
<0.5	8–20	~0–0.5		Spitzen (1987)
				Nobrega
0.1	55–90	1–30		(1988)
~0.67	70–380	0.03–0.7		Huang (1990)
~0.5	80–140	0.003–0.1		Kang (1990a)
~0.5	80–145	0.01–1.0	Surface treated with monochloroacetic acid	Kang (1990a)

Table 3   Ethanol/Water Pervaporation Data: Cellulose-Based Membranes (Water-Selective)

Membrane material	Type of data	Ethanol concentration in feed (wt %)	Temperature (°C)
Cellulose acetate	J_T vs. X_E	0–100	
	α vs. X_E	4–92	
	Y_E vs. X_E	22–96	25
	Z_E, Y_E vs. θ		
	J_E vs. X_E	0–100	
Cellulose triacetate	J_T vs. T	25	
(homogeneous)	J_T vs. X_E	0–95	20–50
	α vs. X_E	5–90	
Cellulose triacetate (homogeneous)	J_E, J_W vs. X_E		0–100
	P_e vs. X_E	0–100	20–50
	Y_W vs. X_W	5–95	
	α vs. X_W	10–95	
Cellulose triacetate (asymmetric)	J_T vs. X_E	0–100	20
Cellulose triacetate	J_T vs. X_W	10–97	
	J_W vs. X_W	0–100	
	J_E vs. X_W	0–100	60
	α vs. X_W	15–97	
	Y_W vs. X_W	10–97	
Cellulose acetate/cellulose triacetate	J_T vs. X_E	5–95	
blend (asymmetric)	α vs. X_E	5–95	20
	J_E vs. X_E	0–100	
	J_W vs. X_E	0–100	
	Y_E vs. X_E	15–90	
	P_e vs. X_W	15–90	
Ethylcellulose	α	7	
Cellulose nitrate/poly(methyl	J_T vs. X_E	0–100	
acrylate)	α vs. X_E	10–90	30
	J_W, J_E vs. X_E		0–100
Carboxymethylcellulose/cellulose	PR, α	80	25
sulfate (sodium salt)	α vs. DOS		
Carboxymethylcellulose	PR vs. X_W	75–95	25
	α	80	

Permeate pressure (kPa)	Separation factor (α_{WE})	Flux (kg/m²·hr)	Comments	Ref.
0.4	0.1–0.5 5–12		Θ is stage cut. Z_E is ethanol concentration in retentate.	Seok (1987)
~0.01	4–12	0–0.7	J and α vs. X_E fit to equations. α highest at low X_W.	Changlou (1988)
~0.01	3–10	0.16–0.7	J_W passes through maximum at $X_E \sim 20$ wt %. Membrane thickness (25 μm). P_e is permeability	Changlou (1989)
~0.01	NR	0.3–0.6	J vs. X_E fit to equations. Minimum J at $X_E \sim 25$ wt %.	Changlou (1988)
2.0	6–14	0.4–1.4		Wenzlaff (1985)
~0.01	1.0–3.6	0.3–1.2	Composite data for two different membranes of the same type. P_e is permeability.	Changlou (1987)
~0.6	2.5	NR ~0–0.5		Lee (1989) Suzuki
–	<1–∞			(1982)
NR	900	0.4	PR is permeation rate (g·mm/ m²·hr). J_T approximated from approximate membrane thickness (1.0 mm) and PR. DOS is degree of substitution.	Reineke (1987)
NR	500	~0–0.2	PR is permeation rate (g·mm/ m²·hr). J_T approximated from approximate membrane thickness (1.0 mm) and PR.	Reineke (1987)

Table 3 Continued

Membrane material	Type of data	Ethanol concentration in feed (wt %)	Temperature (°C)
Carboxymethylcellulose/polyacrylic acid (Na+ counterion)	PR vs. X_W α vs. X_W	80–90	25
Carboxymethylcellulose/polyacrylic acid (crosslinked) (Na+ counterion)	PR vs. X_W α PR vs. T	76–100 80	25 25–75
Carboxymethylcellulose/poly(sodium vinyl sulfonate)	PR vs. X_W α vs. X_W	81–95	25
Carboxymethylcellulose/polyacrylic (Cs+ counterion)	PR vs. X_W J_T, α	82–100 90	 25

Table 4 Ethanol/Water Pervaporation Data: Polysulfone Membranes (Water-Selective)

Membrane material	Type of data	Ethanol concentration in feed (wt %)	Temperature (°C)
Polysulfone	α	7	37
Polysulfone (asymmetric)	J_T vs. X_E α vs. X_E	0–100 5–95	20
Polysulfone (asymmetric)	J_T vs. X_E α vs. X_E Y_E vs. X_E J_E vs. X_E J_W vs. X_E P_e vs. X_W	5–95 5–95 15–95 0–100 0–100 0–100	 20
Polysulfone (asymmetric)	Y_W vs. X_W α vs. X_W	5–95	20–50
Polysulfone on Dacron (composite)	J_T vs. X_E α vs. X_E Y_E vs. X_E J_W vs. X_E J_E vs. X_E P_e vs. X_E	5–95 5–95 15–95 0–100 0–100 0–100	 20

Permeate pressure (kPa)	Separation factor (α_{WE})	Flux (kg/m^2·hr)	Comments	Ref.
NR	610–4100	0.05–0.4	PR is permeation rate (g·mm/ m^2·hr). J_T approximated from approximate membrane thickness (1.0 mm) and PR.	Reineke (1987)
NR	~4000	~0–0.3	PR is permeation rate (g·mm/ m^2·hr). J_T approximated from approximate membrane thickness (1.0 mm) and PR.	Reineke (1987)
NR	2400– 5900	0.005– 0.1	PR is permeation rate (g·mm/ m^2·hr). J_T approximated from approximate membrane thickness (1.0 mm) and PR.	Reineke (1987)
NR	2000	~0–0.6	PR is permeation rate (g·mm/ m^2·hr). J_T approximated from approximate membrane thickness (1.0 μm) and PR.	Reineke (1987)

Permeate pressure (kPa)	Separation factor (α_{WE})	Flux (kg/m^2·hr)	Comments	Ref.
~0.7	3.3	NR	Membranes designated NMP and THF.	Lee (1989)
~0.01	3–7	0.01–0.05	J_T and α vs. X_E fit to equations.	Changlou (1988)
~0.01	3–8	0.02–0.12	Composite data for two different membranes of same type. P_e is permeability.	Changlou (1987)
~0.01	3–6	NR	Membrane thickness 150 μm.	Changlou (1989)
~0.01	2–7	0.03–0.35	P_e is permeability.	Changlou (1987)

Table 5 Ethanol/Water Pervaporation Data: Polyacrylic Acid Membranes (Water-Selective)

Membrane material	Type of data	Ethanol concentration in feed (wt %)	Temperature (°C)
Acrylic acid	J_T vs. X_E	0–90	40
	α vs. X_E	20–90	
Acrylic acid (Na+ counterion)	J_T vs. X_E	0–90	40
	α vs. X_E	20–90	
Acrylic acid on low-density polyethylene	J_T vs. X_W	30–100	35
	Y_W vs. X_W	0–100	
	J_T	>90	
	J_T vs. T		
Acrylic acid on polypropylene GPC	J_T vs. X_E	0–90	
	α vs. X_E	20–90	40
	Y_E vs. X_E	20–90	
Acrylic acid on polypropylene GPC (Na+ counterion)	J_T vs. X_E	0–90	
	α vs. X_E	20–90	40
	Y_E vs. X_E	20–90	
Acrylic acid/acrylamide GPC	J_T vs. X_E	0–90	40
	α vs. X_E		
Acrylic acid/acrylamide GPC (Na+ counterion)	J_T vs. X_E	0–90	40
	α vs. X_E	20–90	
Acrylic acid/2-hydroxyethyl methacrylate	J_T vs. X_E	0–90	40
	α vs. X_E	20–90	
	J_T vs. T		20–60
Acrylic acid/HEMA (Na+ counterion)	J_T vs. X_E	0–90	40
	α vs. X_E	10–90	
Polyacrylic acid/diepoxide	J_T, α	95	70
Polyacrylic acid/diamine (K+ counterion)	J_T, α	95	70
Polyacrylic acid–polycation (PCA-107)	J_T vs. X_E	20–100	70
	α vs. X_E		
Polyacrylic acid/polyallylmine	J_T vs. X_E	95	70
	α vs. X_E		

Permeate pressure (kPa)	Separation factor (α_{WE})	Flux (kg/m²·hr)	Comments	Ref.
<0.1	1–5	~4		Hirotsu (1988b)
<0.1	20–75	0.5–10		Hirotsu (1988b)
Vac	Moderate	0.1–5	Lowest flux for X_E > 90 wt %. Vac means PV test performed	Neel (1987)
		0.1	under vacuum; exact permeate side pressure not stated. Additional cations studied. Na has best performance.	Ping (1990)
<0.1	2–4	2–3.5	GPC is graft-polymerized copolymer.	Hirotsu (1987a)
<0.1	13–80	~0–8	GPC is graft-polymerized copolymer.	Hirotsu (1987a)
<0.1	4–10	0–1.5	Acrylic-acid-to-acrylate (AA:A) ratio is 1:1. GPC is graft-polymerized copolymer.	Hirotsu (1988a)
<0.1	1–20	0–10	AA:A ratio in membrane is 1:1. GPC is graft-polymerized copolymer.	Hirotsu (1988b)
<0.1	5–35	0.2–1.4	Composite data for different membranes of same type. Membrane contains 20 wt % acrylic acid. Performance depends on polypropylene support material.	Hirotsu (1989)
<0.1	3–18	0.3	Composite data for different membranes of same type. Membrane contains 20 wt % acrylic acid. HEMA is 2-hydroxyethyl methacrylate.	Hirotsu (1989)
NR	2100	0.96	Unstable: K^+ ions leach out of membrane in 3 hr.	Karakane (1988)
NR	840	1.06	Unstable: K^+ ions leach out of membrane in 3 hr.	Karakane (1988)
NR	<1–2000	0.5–20	J_T (X_E = 95 wt %) equals 0.8 KMH. α (X_E = 95 wt %) equals 2000. Ethanol selective at X_E < 40 wt %.	Karakane (1988)
NR	~750	~0.5		Karakane (1988)

Table 6 Ethanol/Water Pervaporation Data: Miscellaenous Acrylate-Based Membranes (Water-Selective)

Membrane material	Type of data	Ethanol concentration in feed (wt %)	Temperature (°C)
Methacrylate on polypropylene GPC	J_T vs. X_E	0–90	40
	α vs. X_E		
Methacrylic acid on polypropylene GPC	J_T vs. X_E	0–90	40
	α vs. X_E	15–90	
	Y_E vs. X_E	15–90	
Methacrylate polypropylene GPC (Na$^+$ counterion)	J_T vs. X_E	0–90	40
	α vs. X_E		
Methacrylic acid/2-hydroxyethyl methacrylate (HEMA)	J_T vs. X_E	0–90	40
	α vs. X_E	20–90	
Methacrylic acid/HEMA (Na$^+$ counterion)	J_T vs. X_E	0–90	40
	α vs. X_E	20–190	
HEMA on polypropylene GPC	J_T vs. X_E	20–90	40
	α vs. X_E	10–90	
HEMA on polypropylene GPC	J_T vs. X_E	0–90	40
	α vs. X_E	10–90	
Acrylamide on polypropylene GPC	J_T vs. X_E	0–90	40
	α vs. X_E		
N-Vinylpyrrolidone (NVP)/isobutyl methacrylate	α vs. X_E	10–80	25
4-Vinylpyridine/methyl methacrylate	α vs. X_E	20–75	25

specific membrane surface area of 150 m^2/m^3, is described by Sander and Soukup (1988a); they also discuss many practical aspects of PV system engineering.

The first pilot/commercial scale PV plant for dehydration of ethanol has been operating in Germany since September 1984 (Sander and Soukup, 1988a). This plant produces ethanol from waste sulfite liquor. It uses the GFT membrane and was designed by Lurgi. The PV unit produces 6.0 m^3 anhydrous (99.7–99.9 wt %) ethanol/day from a feed containing 92–94 wt % ethanol. The ethanol produced at this plant does not contain entrainer and is of higher quality than the

Permeate pressure (kPa)	Separation factor (α_{WE})	Flux (kg/m²·hr)	Comments	Ref.
<0.1	2–5	0.4–1.5	GPC is graft-polymerized copolymer.	Hirotsu (1987a)
<0.1	1–20	~0–8	GPC is graft-polymerized copolymer.	Hirotsu (1987b)
<0.1	2–29	0.5–7.8	GPC is graft-polymerized copolymer.	Hirostu (1987a)
<0.1	4–40	0.05–0.4	Membrane contains 20 wt % methacrylic acid. Performance depends on polypropylene support material.	Hirotsu (1989)
<0.1	4–50	0.2–5.4	Membrane contains 20 wt % methacrylic acid. Performance depends on polypropylene support material.	Hirotsu (1989)
<0.1	1–60	~0–7	Highest J_T occurs at $\alpha \sim 1$. GPC is graft-polymerized copolymer.	Hirotsu (1988a)
<0.1	15–65	0.2–0.6	Performance depends on polypropylene support material. Combined data for different membranes of same type.	Hirotsu (1989)
<0.1	1.4–11	0.8–3.5	GPC is graft-polymerized copolymer.	Hirotsu (1987a)
NR	1.5–12	NR	Membrane thicknesses: 43–52 μm. Membranes contained 7–21% NVP.	Yamada (1987)
NR	2–41	NR	Membrane thicknesses: 62–74 μm.	Yamada (1987)

product produced by azeotropic distillation. Based on this operation experience, membrane life is expected to be 2–4 years.

Since the first commercial plant in Germany, GFT has sold more than 100 plants around the world for dehydration of organic chemicals. The largest plant has been operated since 1988 at Betheniville, France (Rapin, 1988). This plant produces ethanol from sugarbeets or wine molasses. The GFT PV unit produces 150 m³ anhydrous (99.8–99.95 wt %) ethanol/day from a feedstream containing ethanol near azeotropic composition. This performance corresponds to an ethanol-

Table 7 Ethanol/Water Pervaporation Data: Polyvinyl Fluoride Membranes (Water-Selective)

Membrane material	Type of data	Ethanol concentration in feed (wt %)	Temperature (°C)	Permeate pressure (kPa)	Separation factor (α_{WE})	Flux (kg/m²·hr)	Comments	Ref.
Polyvinyl fluoride/4-vinylpyrrolidone (w/betaine)	J_T α	80	70	NR	~60	1.25	Membrane reported to be stable after 50 days of pervaporation use.	Ellinghorst (1987)
Polyvinyl fluoride/N-vinyl methyl acetemide GPC	J, α	80	70	~3.0	4	2.9	GPC is graft-polymerized copolymer.	Niemoller (1988)
Polyvinyl fluoride/acrylic acid	Y_E vs. X_E	17–95	67	NR	Low to moderate	NR	Membrane reported to be unstable.	Ellinghorst (1987)
Polyvinyl fluoride/acrylic acid (Na⁺ counterion)	J_T	80	70	NR	Moderate	1.8	Membrane reported to be unstable.	Ellinghorst (1987)
Polyvinyl fluoride/acrylic acid GPC (K⁺ counterion)	J, α; Y_E vs. X_E	80; 5–90	70	3.0	250 High	2.0	GPC is graft-polymerized copolymer.	Niemoller (1988)
Polyvinyl fluoride/acrylic acid (K⁺ counterion)	J_T; Y_E vs. X_E	80; 40–95	70; 67	NR	High	4.7	Membrane reported to be unstable.	Ellinghorst (1987)
Polyvinyl fluoride/N-vinyl methyl acetemide	J_T α	80	70	NR	4	2.9	Membrane reported to be unstable.	Ellinghorst (1987)
Polyvinyl fluoride/N-vinylpyrrolidone	J_T α	80	70	NR	7	10	Membrane reported to be unstable.	Ellinghorst (1987)

to-water separation factor of 25–200, depending on the actual feed and product concentrations. The permeate flow rate at this plant is 430 m³/day at 20 wt % ethanol; the permeate is recycled to an ethanol rectification column.

In addition, two plants in Japan are using GFT/Mitsui PV technology to dehydrate ethanol (Asada, 1988). At one plant, the production rate is 1.5 m³ ethanol (99.7 wt %)/day (50 kg/hr) from a feed containing 67 wt % ethanol. The second plant produces about 40 m³ ethanol (96 wt %)/day (1300 kg/hr) from a feed containing 90 wt % ethanol.

The energy requirements of a hybrid distillation/PV plant are about 3.7 GJ/m³, which compares favorably to conventional distillation (6.7–8.2 GJ/m³) (Serra, 1987; Table 1). Vapor recompression coupled to azeotropic distillation with ether has lower energy requirements (2.5 GJ/m³) but higher capital costs. PV compares more favorably with advanced azeotropic distillation if the feed concentration to the dehydration step is higher. For instance, Cogat (1988) reports that PV requires about 0.4 GJ/m³ (incremental energy requirement for dehydration from about 95/96 to 99.9 wt %), whereas azeotropic distillation requires about 1.1–3.3 GJ/m³ for the same incremental purification. Fleming (1989a) reports an incremental energy requirement of about 0.6 GJ/m³ for PV and 3.1 GJ/m³ for conventional azeotropic distillation; the feed concentration is not reported, but the product concentration is 99.9 wt %. According to Changlou et al. (1987), PV requires about 0.3 GJ/m³ to dehydrate 95 wt % ethanol to 99 wt % ethanol.

The costs of PV are provided by Fleming (1989a). For the PV unit, the costs range from $12 to $42/liter/day for ethanol capacities of 180–1.8 m³/day. Entire plant costs range from $83 to $167/liter/day for ethanol capacities of 12–2.4 m³/day. According to Gooding (1986), the incremental operating costs (including the cost of capital) of ethanol dehydration is about $0.11/kg.

Ethanol has been dehydrated at the laboratory scale using continuous membrane column pervaporation. This concept was first described for RO by Loeb and Bloch (1973) and for PV by Hwang and Thorman (1980). In summary, this concept involves recycling a fraction of the product stream on the permeate side of the membrane to reduce mass transfer resistances or to change the separation characteristics. Hoover and Hwang (1982) concentrated 94 wt % ethanol to 96.3 wt % ethanol using silicone rubber PV membranes, which indicates that this concept can be used to break the ethanol-water azeotrope.

Another approach to the use of PV for production of anhydrous ethanol is the two column continuous concept (Seok et al., 1987). As this concept requires the use of water-selective and ethanol-selective membranes, it is described at the end of the next section.

Pervaporation: Ethanol Concentration (Ethanol-Selective Membranes)

PV is receiving intense study as a means to concentrate dilute ethanol and to directly recover ethanol from fermentation broth. Various ethanol-selective mem-

Table 8 Ethanol/Water Pervaporation Data: Polyvinyl-Based Membranes (Water-Selective)

Membrane material	Type of data	Ethanol concentration in feed (wt %)	Temperature (°C)
Polyvinylidene fluoride/N-vinylpyrrolidone	Y_E vs. X_E	5–95	67
Polyvinylidene fluoride/N-vinylpyrrolidone GPC	J, α Y_E vs. X_E	80 5–90	70
Polyvinylidene fluoride/4-vinylpyridine	J_T, α	80	70
Polyvinylidene fluoride/4-vinylpyridine GPC	J, α	80	70
Polyvinylidene fluoride/4-vinylpyridine (quaternized) GPC	J, α Y_E vs. X_E	80 35–95	70
Polyvinylidene fluoride/4-vinylpyridine (w/betaine functions)	J_T, α	80	70
Polyvinylidene fluoride/4-vinylpyridine (methylated)	J_T Y_E vs. X_E	80 35–95	70 67
Polyvinylidene fluoride/N-vinylimidazole	J_T Y_E vs. X_E	80 35–95	70 67
Polyvinylidene fluoride/N-vinylimidazole GPC	J_T vs. X_E Y_E vs. X_E	0–95 35–92	70
Polyvinylidene fluoride/N-vinylimidazole (quaternized) GPC	J_T vs. X_E Y_E vs. X_E	0–95 35–92	70
Polyvinylidene fluoride/N-vinylimidazole (methylated)	J_T Y_E vs. X_E	80 35–95	70 67
Polyvinylidene fluoride/acrylic acid (K+ counterion)	Y_E vs. X_E	35–95	67
Polyvinylidene fluoride w/radiation-cured coating	J_T, α	80	70

branes are discussed, followed by a discussion of proposed processes, experimental results, energy requirements, and cost estimates.

Silicon rubber (polydimethylsiloxane) and other silicon-based polymers are the most thoroughly studied ethanol-selective membranes. The emphasis of this chapter is membrane applications for recovery of ethanol produced by fermentation. A discussion of the use of silicone rubber PV membranes as an approach to reducing the alcohol content of beer and wine is provided by Bruschke (1990b). A theoretical discussion of ethanol-water transport in silicone rubber membranes is provided by Radovanovic et al. (1990) and by Watson and Payne (1990).

Permeate pressure (kPa)	Separation factor (α_{WE})	Flux (kg/m²·hr)	Comments	Ref.
NR	Low	NR	Membrane is reported to be unstable.	Ellinghorst (1987)
~3.0	7 Low to moderate	0.8	GPC is graft-polymerized copolymer.	Niemoller (1988)
NR	9	0.2		Ellinghorst (1987)
~3.0	8	0.67	GPC is graft-polymerized copolymer.	Niemoller (1988)
3.0	73 Moderate to high	1.0	GPC is graft-polymerized copolymer.	Niemoller (1988)
NR	~76	0.2	Membrane is reported to be stable.	Ellinghorst (1987)
NR	Moderate to high	0.4		Ellinghorst (1987)
NR	Moderate	1.6	Membrane is reported to be unstable.	Ellinghorst (1987)
3.0	Moderate	0.2–5	GPC is graft-polymerized copolymer.	Niemoller (1988)
3.0	Moderate	~0–6	GPC is graft-polymerized copolymer.	Niemoller (1988)
NR	Moderate to high	0.4	Membrane is reported to be unstable.	Ellinghorst (1987)
NR	High	NR	Membrane is reported to be unstable.	Ellinghorst (1987)
NR	326	3.9	Membrane designated PVF$_2$-V-1; no details provided about coating.	Ellinghorst (1987)

Ethanol/water PV data for silicon rubber and silicon rubber copolymer membranes are presented in Table 12. These membranes are characterized by low to moderate separation factors (α_{EW}) and generally exhibit low flux. Since silicone rubber membranes perform the greater degree of separation at low ethanol concentration, they may be useful for removing dilute ethanol from water; however, flux is also lowest at low ethanol concentrations.

The search for ethanol-selective membranes with high separation factors and flux has led to the study of a wide range of silicon-based materials (Table 13). Various silane and siloxane polymers exhibit low, but potentially useful, separa-

Table 9 Ethanol/Water Pervaporation Data: Miscellaneous Fluorinated Membranes
(Water-Selective)

Membrane material	Type of data	Ethanol concentration in feed (wt %)	Temperature (°C)	Permeate pressure (kPA)
Nafion (K$^+$ counterion)	J_T vs. X_E	30–98		
	Y_E vs. X_E	15–90	40	<0.1
Nafion (K$^+$, Cs$^+$)	J_T, α	22	30	<0.1
Polyhexafluoropropene-*co*-tetrafluoroethene/vinyl acetate GPC	J, α	80	70	~3.0
Polyhexafluoropropene-*co*-tetrafluoroethene/vinyl acetate GPC	J, α	80	70	3.0
Polyhydroxymethylene-*co*-fluoroolefin	J_T vs. X_E α vs. X_E J_T α vs. T	50–95	25–75	0.01

tion factors and reasonable flux. Silicon-based oils dissolved in polypropylene
exhibit moderate separation factors but low flux. Fluorinated silicon-based membranes exhibit low separation factors.

Membranes made from fluorine-containing polymers are ethanol-selective
(Table 14). Polytetrafluoroethylene on polypropylene exhibits high flux and low to
moderate separation factors; these membranes may provide viable alternatives to
silicone rubber membranes. Hexafluoroethane allylamine membranes exhibit high
flux with low selectivity and high selectivity with low flux.

PV data are reported for miscellaneous and unidentified membranes in Table
15. Since many of these data are incomplete, it is difficult to draw conclusions.
However, the poly[1-(trimethylsilyl)-1-propyne] (PTMSP) membrane is worthy of
note; preliminary data on this membrane indicate that it has performance equivalent or superior to that of silicone rubber.

A PV-based process for direct ethanol recovery from fermentation broth or for
recovery of ethanol from dilute aqueous solutions might look like the schematic
presented in Figure 2. The retentate stream, which is depleted in ethanol, is
recycled to the fermentor; if this stream has a low ethanol concentration, it should
not adversely affect the fermentation and could provide a source of process water.
The feed to the PV unit may require heating in order to optimize the process, since
flux increases with increasing temperature. If the feed needs to be heated, then the

Separation factor (α_{WE})	Flux (kg/m^2·hr)	Comments	Ref.
Moderate to low	~0–0.5	α lowest at high X_E. Nafion is the trade name for perfluorosulfonic acid on polytetrafluoroethylene. Used Nafion 117 (DuPont).	Kujawski (1988)
8–10	0.18–0.26	Nafion 811 hollow fibers (DuPont) with a wall thickness of 90 μm.	Cabasso (1985a)
4	1.6	Membrane reported by Niemoller (1988) to be unstable. GPC is graft-polymerized copolymer.	Niemoller (1988) Ellinghorst (1987)
56	2.8	Membrane reported by Niemoller (1988) to be unstable. GPC is graft-polymerized copolymer.	Niemoller (1988) Ellinghorst (1987)
900–2400	~0–0.25		Terada (1988)

recycled retentate stream may require cooling (unless the fermentation can be operated at thermophilic temperatures) (Bitter, 1988).

Polytetrafluoroethylene (PTFE) hollow-fiber modules have been studied for ethanol stripping (Calibo et al., 1987, 1989). The fundamental performance of PTFE was characterized using aqueous ethanol feedstreams (Calibo et al., 1987). Then PTFE hollow-fiber modules were studied to remove ethanol from fermentation broth; the fermentation of glucose and molasses with yeast was performed at 30°C and the PV experiments were performed at 40°C. Product yield and volumetric productivity were increased by removing ethanol through the PTFE membranes by PV. In addition, PV module performance did not change during long-term runs (Calibo et al., 1989).

The PV bioreactor concept has also been studied by Nakao et al. (1987) using a PTFE membrane. They report a three- to fourfold increase in volumetric productivity when operating in a partial cell recycle mode. Removal of cells was necessary to maximize productivity.

Mori and Inaba (1990) studied PTMSP and polytetrafluorethylene/silicone rubber (PTFE/SR) membranes for use in a PV bioreactor. The performance of the PTMSP membrane was dramatically reduced by contact with fermentation broth (relative to its performance with solutions of pure ethanol and water), whereas the PTFE/SR membrane was not. Therefore, the PTFE/SR membrane was studied in

Table 10 Ethanol/Water Pervaporation Data: Polyacrylonitrile Membranes (Water-Selective)

Membrane material	Type of data	Ethanol concentration in feed (wt %)	Temperature (°C)
Polyacrylonitrile (plasma grafted)	J_T, α	80	65
Polyacrylonitrile/polyvinylpyrrolidone	J_T vs. X_W β vs. X_W α	78–98 78–98 95.6	20
Polyacrylonitrile/acrylic acid GPC	J, α	80	70
Polyacrylonitrile/acrylic acid GPC	J_T, α	80	70
Polyacrylonitrile/acrylic acid GPC (Na$^+$ counterion)	J_T, α	80	70
Polyacrylonitrile/acrylic acid GPC (K$^+$ counterion)	J, α	80	70
Polymaleimide-*co*-acrylonitrile	J_T vs. X_W α vs. X_W Y_W vs. X_W α vs. T	10–95	15 −10–40
Polyacrylonitrile w/radiation-cured coating	J_T, α	80	70
Polyacrylonitrile w/radiation-cured coating	J_T, α	80	70

Table 11 Ethanol/Water Pervaporation Data: Miscellaneous Membranes (Water-Selective)

Membrane material	Type of data	Ethanol concentration in feed (wt %)	Temperature (°C)
Polyethylene-*co*-styrenesulfonate	J_T vs. X_E Y_W vs. X_W	5–70 5–95	40
Polyvinyl acetate on polysulfone (thin-film composite)	α vs. X_W Y_W vs. X_W	5–95 0–100	20–50
Polyvinyl acetate on polysulfone (composite)	J_T vs. X_E α vs. X_E	5–95	40
Polyphenylquinoxaline	J_T vs. X_E α vs. X_E	10–85	70

Permeate pressure (kPa)	Separation factor (α_{WE})	Flux (kg/m²·hr)	Comments	Ref.
NR	400–4000	0.4		Ellinghorst (1987)
~0.1		1.5–2.5	β is a separation performance (defined Y_W/X_W).	Nguyen (1985)
	3.2			
~3.0	9	1.6	GPC is graft polymerized copolymer.	Niemoller (1988)
NR	10	1.8	Membrane is reported to be unstable. GPC is graft-polymerized copolymer.	Ellinghorst (1987)
NR	~1000	~2.4	Membrane is reported to be unstable. GPC is graft-polymerized copolymer.	Ellinghorst (1987)
~3.0	866	3.4	GPC is graft-polymerized copolymer.	Niemoller (1988)
			α highest at high X_E. α decreases with increasing T.	Yoshikawa (1984)
~0.37	30–350	~0–0.08		
NR	~800	1.7	Designated PAN-C-1; no details provided on coating.	Ellinghorst (1987)
NR	~440	4.2	Designated PAN-C-2; no details provided on coating.	Ellinghorst (1987)

Permeate pressure (kPa)	Separation factor (α_{WE})	Flux (kg/m²·hr)	Comments	Ref.
<0.1	High to moderate	~0–0.4	α highest at low X_E.	Kujawski (1988)
0.01	5–100	NR	Membrane thickness (5 μm).	Changlou (1989)
~0.01	2–100	0–0.01	J_T and α vs. X_E fit to equations.	Changlou (1988)
0.1	2–6	~0.20		Guanwen (1988)

Table 11 Continued

Membrane material	Type of data	Ethanol concentration in feed (wt %)	Temperature (°C)
Polyphenylquinoxaline (w/ ether linkages)	J_T vs. X_E α vs. X_E	10–85	70
Polyhydroxymethylene	J_T vs. T α vs. T	95	25–75
Polybenzimidazole	J_T vs. X_E Y_W vs. X_W	2.5–99.6	30
Polyetheramide (UOP)	J vs. X_E Y_W vs. X_E	12–58 65–95	35–45
Polytetramethylene glycol-block-nylon-12	α	1–5	NR
Styrene/butadiene crosslinked on polyvinyl chloride (Na$^+$ counterion)	J_T vs. X_W J_W, J_E vs. X_W α vs. X_W Y_W vs. X_W	15–80 0–100 10–95 15–95	60
Styrene/butadiene crosslinked on polyvinyl chloride (Mg^{2+} counterion)	J_T vs. X_W J_W, J_E vs. X_W α vs. X_W Y_W vs. X_W α vs. X_W	15–95 0–100 15–95 10–99 55–80	60
Styrene/butadiene crosslinked on polyvinyl chloride (SO$_4^{2-}$ counteranion)	J_T vs. X_W J_W, J_E vs. X_W α vs. X_W Y_W vs. X_W α vs. X_W	15–98 0–100 15–99 10–99 96–99	60
Chitosan	Y_E vs. X_E	0–100	25
Nafion (Cs$^+$ counterion)	J_T vs. T α vs. T	78	24–40
Sulfonated polyethylene (Cs$^+$ counterion)	J_T, α	85.4	26
Polyurethane/MAA/AA	P vs. X_E α vs. X_E Y_E vs. X_E	0–100	30
Poly(1-phenyl-1-propyne)	Y_E vs. X_E	30–90	30

Permeate pressure (kPa)	Separation factor (α_{WE})	Flux (kg/m²·hr)	Comments	Ref.
0.1	2–12	0.05–0.1		Guanwen (1988)
0.01	>4900	~0–0.01		Terada (1988)
NR	Moderate to low	0.2–0.4	Maximum J_T at ~ 15 wt % ethanol.	Sferrazza (1988)
~1.4	Low	1.7–2.1	UOP Fluid Systems (San Diego, CA) membrane.	Gooding (1986)
NR	1.1	NAR	NAR: not adequately reported. Membrane designated 5533, with 67% nylon content. At <50% nylon content, these membranes were ethanol-selective.	Morin (1988)
2.0	2–8	2.0–2.6	α is approximately constant at 8 for X_E equals 10–45 wt %.	Wenzlaff (1985)
2.0	2–11	0.8–1.7		Wenzlaff (1985)
2.0	5–70	8–11 / 0.1–1.2		Wenzlaff (1985)
NR	>55 / Comments	NR	Water-selective (for X_E > 20 wt %) with moderate to high α. Ethanol-selective for X_E < 17.5 wt % with very low α.	Uragami (1988)
<0.1	~16	0.4–0.8	Additional cations studied (K, Na, Li, H). Hollow-fiber membranes with thickness of 90μm.	Cabasso (1985a)
<0.1	725	0.152	Additional cations studied (K, Na, Li, H). Membrane thickness is 40.3 μm	Cabasso (1985b)
0.01	<1–10	NR	P is permeability in g·cm/cm²·hr. Membrane thickness not reported. MAA is methylacrylic acid and AA is acrylic acid.	Lee (1990)
NR	Low	NR		Nagase (1987)

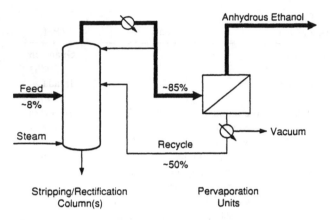

Figure 1 Process for dehydration of ethanol by pervaporation.

Table 12 Ethanol/Water Pervaporation Data: Silicone Rubber Membranes (Ethanol-Selective)

Membrane material	Type of data	Ethanol concentration in feed (wt %)	Temperature (°C)
Silicone rubber on polysulfone	J_T α	43.5	40
Silicone rubber	J_T vs. X_E Y_E vs. X_E α vs. X_E	0–50	30
PDMS (homogeneous)	J_T, α J_T, α	5 95	25
Silicone rubber	α vs. X_E	0–8	NR
SIlicone rubber	J_T vs. X_E α vs. X_E J_T vs. T J_T, α vs. PP	3–100 3–88 21	25 25 21–50 0.07–6.7
Silicone rubber	α	7	37
Silicone rubber	J_T, α	15	66
Silicone rubber	J_T vs. X_E α vs. X_W Y_E vs. X_E Z_E, Y_E vs. Θ	22–96	25
Silicone rubber	J_W vs. X_E J_E vs. X_E	0–100	30

continuous fermentations. Average flux was ~0.8 KMH, with an overall separation factor of 8.3. The volumetric productivity of the PV bioreactor was 0.36 g/liter·hr, which was 2.25 times higher than that of a conventional continuous bioreactor. The authors conclude nonetheless that a higher separation factor is needed before such a process can achieve economic feasibility.

PV can be used to purify dilute ethanol solutions and produce anhydrous ethanol. The entire purification of ethanol can be completed by combining an ethanol-selective PV membrane system and a water-selective PV system.

Bitter (1988) analyzed a PV process for complete purification of ethanol. The ethanol-selective PV unit concentrates ethanol from 5 to 40 wt %. The water-selective PV unit produces 99.5 wt % ethanol. An overall separation factor (α_{EW}) of about 55 was assumed for the first (ethanol-concentrating) PV unit; ethanol-selective membranes with selectivities in that range are not presently available.

Permeate pressure (kPa)	Separation factor (α_{EW})	Flux (kg/m²·hr)	Comments	Ref.
~0.07	4.2	0.71	Silicone rubber is polydimethyl-siloxane (PDMS)	Belanche (1988)
0–7	2–8	0.002	Membrane thickness is 2.2 μm.	Blume (1987) Blume (1990)
~0.01	9	0.001	PDMS is polydimethylsiloxane (silicone rubber).	Changlou (1988)
	~1	0.003		
NR	2–17	NR		Watson (1990)
~0.7	3–14	0.02–17	PP is permeate side pressure.	Kimura (1983)
~9	~0–0.05			
~0.6	8.4	NR		Lee (1989)
5.3	10.4	0.15	Membrane thickness is 120 μm.	Mori (1990)
0.33	1–11	NR	Selectivity (α_{EW}) is moderate for X_E < 46 wt % and low for X_E > 46 wt %. Θ is stage cut. X_E is ethanol concentration in retentate.	Seok (1987)
~0.2	NR	~0–2.75		Radovanovic (1988)

Table 12   Continued

Membrane material	Type of data	Ethanol concentration in feed (wt %)	Temperature (°C)
Silicone rubber	J_T vs. X_E	0–100	25
	α vs. X_E	20–90	
	Y_E vs. X_E	5–90	
Silicone rubber (modified)	Y_E vs. X_E	0.5–14	NR
Silicone rubber	Y_E vs. X_E	5–10	30
	J_T		
PDMS/polyvinyl fluoride (composite)	J_T vs. X_E	5–95	20
	α vs. X_E	5–95	
	Y_E vs. X_E	5–90	
	J_W, J_E vs. X_E	0–100	
	P vs. X_E	0–100	
PDMS/polyvinyl fluoride (composite)	J_T, α	10–95	NR
Silicone rubber on polyimide	α	7	37
PDMS/polycarbonate (plasma-polymerized)	J_T	4–27	25
	α		
Silicalite-filled silicone rubber	J_T, α	5	25
	J_T vs. X_Z		
Polyolefin/PDMS	Y_E vs. X_E	0–1.0	30
		5–99	
Zeolite-filled silicone rubber	J_T, α	15	66
PTFE/silicone rubber	J_T, α	15	66
PPP/PDMS graft copolymer	Y_E vs. X_E	10–90	30
	α	7	
PTMSP/PDMS graft copolymer	Y_E vs. X_E	10–90	30
PDMS composite on PVDF	J_T, α	8	25
PDMS-PS-PHS composite of PVDF	J_T, α	8	25

Permeate pressure (kPa)	Separation factor (α_{EW})	Flux (kg/m²·hr)	Comments	Ref.
NR	~1–8	~0–0.075		Tanigaki (1987)
NR	Moderate	NR	Modification not described.	Morigami (1987)
~1.0	Moderate	0.04		Nakao (1987)
~0.01	1.2–7	0.07–0.4	PDMS is polydimethylsiloxane (silicone rubber).	Changlou (1987)
NR	1.1–6	0.02–0.2	Selectivity is very low (<2) for $X_E > 60$ wt %. PDMS is polydimethylsiloxane (silicone rubber).	Changlou (1988)
~0.6	1.1	NR		Lee (1989)
<0.1	~1–4.5	~0–4	Reasonable selectivity ($\alpha > 1$) only for $X_E < 15$ wt %. Flux is lowest at low X_E.	Inagagaki (1988)
<0.1	40	~0.001	X_Z is zeolite content of membrane. Membrane for which data are presented contained 70 wt % silicate.	teHennepe (1987)
~1.0	Moderate Low	NR	Membrane thickness: 3.5 μm.	Blume (1990)
5.3	11.1	0.12	Membrane thickness is 200 μm.	Mori (1990)
5.3	14	1.53	PTFE is polytetrafluoroethylene. Membrane thickness is 5 μm.	Mori (1990)
NR	Moderate 40	NR	PPP is poly(1-phenyl-1-propyne). PDMS is polydimethyl-siloxane (silicone rubber).	Nagase (1987)
NR	Moderate	NR	PTMSP is poly[1-(trimethyl-silyl)-1-propyne]. PDMS is polydimethylsiloxane (silicone rubber).	Nagase (1987)
0.13	2–9	0.1–20	PVDF is polyvinylidene fluoride. Membrane thickness is 0.22–0.45 μm.	Okamoto (1987)
0.13	5–7	3–5	PS-PHS is polysulfone/poly(4-hydroxystyrene)	Okamoto (1987)

Table 13 Ethanol/Water Pervaporation Data: Miscellaneous Silicone-Based Membranes (Ethanol-Selective)

Membrane material	Type of data	Ethanol concentration in feed (wt %)	Temperature (°C)
Methoxysilane and derivatives	J_T, α	4	25
Polydimethylsiloxane/octadecyl-diethoxymethylsilane GPC	J_T, α	4	25
Hexamethyltrisiloxane/octadecyl-diethoxymethylsilane GPC	J_T, α	4	25
Hexamethyltrisiloxane (plasma-polymerized)	J_T, α	4	25
Octadecyldimethylmethoxysilane and derivatives	J_T, α	4	25
Methyltrioctylsilane	J_T, α	4	25
Polyvinyldimethylsiloxane	α	7	37
Fluorinated decyltriethoxysilane in porous glass	J_T vs. X_E α vs. X_E	10–90	25
Fluorinated propyltriethoxysilane in porous glass	J_T vs. X_E α vs. X_E	10–90	25
Chlorinated propyltriethoxysilane in porous glass	J_T vs. X_E α vs. X_E	10–90	25
Octyltriethoxysilane in porous glass	J_T vs. X_E α vs. X_E	10–90	25
Pentyltriethoxysilane in porous glass	J_T vs. X_E α vs. X_E	10–90	25

According to Bitter (1988), an α_{EW} of greater than 450 would be needed to make the process economical.

As an alternative, the two-column pervaporator of Seok et al. (1987) is shown in Figure 3. In this concept, the retentate stream, rather than the permeate stream, is the product of each column. The columns are operated at high stage cut, which increases retentate (product) purity. The concentration of the more permeable component in the retentate stream decreases with increasing stage cut; overall flux increases because the flux of the less permeable component increases with stage cut. In single columns, in which the permeate is the product, low stage cut must be used to obtain pure (permeated) product. When operating at low stage cut,

Permeate pressure (kPa)	Separation factor (α_{EW})	Flux (kg/m^2·hr)	Comments	Ref.
NR	1–5	0.1–0.4		Kashiwaga (1988)
NR	16.3	0.016	GPC is graft-polymerized copolymer. Membrane thickness is 20 μm.	Kashiwaga (1988)
NR	18	0.015	GPC is graft-polymerized copolymer. Membrane thickness is ~1 μm.	Kashiwaga (1988)
NR	5	0.32	Membrane thickness ~1 μm.	Kashiwaga (1988)
NR	13–15	0–0.02	Membrane thickness is 25 μm. Oil dissolved in polyethylene.	Kashiwaga (1988)
NR	16.9	~0.005	Membrane thickness is 25 μm. Oil dissolved in polyethylene.	Kashiwaga (1988)
~0.6	5.6	NR		Lee (1989)
NR	Low to moderate	~0–0.1	Designated C8F17. Selectivity highest at low X_E.	Lee (1988)
NR	Low	0.01–0.1	Designated CF3.	Lee (1988)
NR	Very low	0.1–0.15	Designated 3C1.	Lee (1988)
NR	Low	0.04–0.2	Designated C8.	Lee (1988)
NR	Low	0.04–0.15	Designated C5.	Lee (1988)

recovery is low and the retentate stream flow rate is large (relative to the permeate flow rate). The retentate stream must be further processed or recycled. Overall separation factor for the two-column pervaporator is a function of feed flow rates, stage cut of each column, and the ratio of membrane areas. For ethanol/water separation with silicon rubber and cellulose acetate membranes, overall separation factors of 11–26 are reported by Seok et al. (1987).

Blume et al. (1990) estimated the costs of using PV for water treatment and dilute organic chemical recovery. For aqueous feedstreams containing organic chemicals at 1–5 wt %, capital and operating costs are estimated to be between $2 and $5/gallon of feed/day and $2 and $5/1000 gallons of feed.

Table 14   Ethanol/Water Pervaporation Data: Fluorine-Containing Membranes (Ethanol-Selective)

Membrane material	Type of data	Ethanol concentration in feed (wt %)	Temperature (°C)
Polytetrafluoroethylene (PTFE)	Y_E vs. X_E	2–10	30
	J_T vs. X_E		
Polytetrafluoroethylene (PTFE) on polypropylene	J_T vs. T	5	40–65
	α vs. P/P		
Polytetrafluoroethylene (Gore-Tex)	J_T	8	
	J_T vs. T	6	43–50
	Y_E vs. X_E	2–30	
	α vs. T	6	
Polyvinylidene fluoride	J_T vs. X_E	0–15	32
	α vs. X_E	2–15	
	α vs. T	7	13–40
Hexafluoroethaneallylamine/ perfluoropropylene copolymer	α vs. J_T	4.8	40
Hexafluoroethaneallylamine/ perfluoropropane copolymer	α vs. J_T	4.8	40
Hexafluoroethaneallylamine/ tetrafluoroethylene copolymer	α vs. J_T	4.8	40
Various PTFE-based copolymers	J_T, α	9–15	25–60

Pervaporation: Ethanol/Organic Separations

The hybrid use of PV and liquid–liquid extraction may also be useful in the recovery of biofuels and biochemicals. A possible process is shown in Figure 4. Therefore, ethanol/organic PV data are also compiled and discussed. Data for ethanol/*n*-hexane, ethanol/*n*-heptane, and ethanol/toluene separations are presented for various membranes in Table 16. Data for the polytetrafluoroethylene/*N*-vinylpyrrolidone membrane and various ethanol/organic mixtures were compiled in Leeper (1986), based on the work of Aptel et al. (1976). Available data indicate that hybrid PV/extraction processes may be useful in the recovery of ethanol from fermentation broth.

Permeate pressure (kPa)	Separation factor (α_{EW})	Flux (kg/m²·hr)	Comments	Ref.
~1.0	Moderate	5.7		Nakao (1987)
<20	2–10	2–7	PTFE layer has thickness of 60 μm. P/P is ratio of feed side pressure to permeate side pressure.	Bandini (1988)
7	Low	5		Hoffman (1987)
~0.6	2–3.4	1.4–2.2		Lee (1987)
NR	~1–7	0.1–1		Masuoka (1988)
NR	~1–8	0.1–8		Masuoka (1988)
NR	2–6	~0–1		Masuoka (1988)
0.01	0.5–7	0–8	High flux occurs at α of 1.0. Flux of most copolymer membranes very low.	Nakamura (1988)

Vapor Permeation

Vapor permeation (VPe) is emerging as a promising operation for dehydration of ethanol and other organics. VPe could, in theory, also be used for concentration of organics from aqueous solutions.

Vapor Permeation: Ethanol Dehydration

VPe has the advantages over azeotropic distillation listed above for PV. VPe may also have potential advantages over PV. VPe is more compatible with fractional distillation; the feed to VPe needs to be a vapor, so the distillate from the fractional distillation does not need to be completely condensed. Heat input is not required

Table 15   Ethanol/Water Pervaporation Data: Miscellaneous Membranes (Ethanol-Selective)

Membrane material	Type of data	Ethanol concentration in feed (wt %)	Temperature (°C)	Permeate pressure (kPa)
Polypropylene	α vs. PP	5	35–50	<20
Polypropylene	Y_E vs. X_E J_T	5–10	30	~1.0
Polytetramethylene glycol-block-nylon 12	Y_E vs. X_E	1–5	NR	NR
Polyolefin	Y_E vs. X_E	0–1.0	30	~1.0
Poly[1-(trimethylsilyl)-1-propyne]	J_T vs. T, X_E α vs. T, X_E Y_E vs. X_E	10 10–85	10–50 30	0.1 0.01–10 0.01
Poly[1-(trimethylsilyl)-1-propyne]	J_T vs. X_E α vs. X_E Y_E vs. X_E	10–85	30	<0.1
Poly[1-(trimethylsilyl)-1-propyne]	J_T, α	1.5	66	5.3
Substituted polyacetylenes	α, J_T J_T vs. X_E α vs. X_E Y_E vs. X_E	10 10–90	30 30	0.1 0.01
Polycyclohexylmeth-acrylate-*co*-styrene via radical copolymerization	Y_E vs. X_E α vs. X_E α vs. PP J_T vs. X_E, PP	10–90	15	0.4–10
Unidentified membrane	Y_E vs. X_E J_T	3–10 10	30 30	~0.7 ~0.7
Unidentified membrane	Y_E vs. X_E	0.5–9	NR	NR

Separation factor (α_{EW})	Flux (kg/m²·hr)	Comments	Ref.
7–10	NR	Polypropylene used in this study was Celgard 2400 (Celanese) with a thickness of 25 μm. PP is permeate side pressure.	Bandini (1988)
Moderate	2		Nakao (1987)
1.7–2.6	NAR	NAR is not adequately reported. Selectivity depends on membrane composition. For 50:50 and less, the resulting membrane is ethanol-selective. Membranes designated 3533 and 4033.	Morin (1988)
Moderate	NR		Blume (1990)
4–12	0.1–0.4	Membrane thickness is ~20 μm.	Masuda (1986)
1–17	0.3–3.5		Masuda (1990)
1–11	NAR	NAR is not adequately reported.	Ishihara (1986)
18.7	1.57	Performance data presented for ethanol-water solution; with fermentation broth, performance significantly lower. Membrane thickness is 30 μm.	Mori (1990)
1.4–4.8	0.01–0.03	Substituted groups include H, CH_3, t-Bu, C_6H_5, C_5H_{11}, C_6H_{13}, Cl, silicones, etc.	Masuda (1986)
1–8	~0.1–0.3		Masuda (1990)
1–5	<0.1	Membrane thickness is 7–10 μm. PP is permeate side pressure.	Yoshikawa (1987)
Low to moderate 7.5	~0.08	MTR Pervap membrane (Code 100).	Kaschemekat (1988)
Moderate to high	NR	Designated new membrane. Selectivity highest at low X_E.	Morigami (1987)

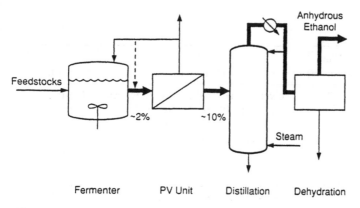

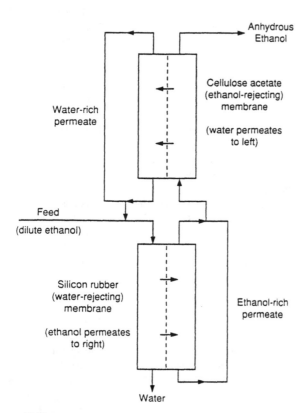

Figure 2 Hybrid process for ethanol recovery by pervaporation and distillation (PV bioreactor concept).

Figure 3 The two-column pervaporator concept (From Seok et al., 1987.)

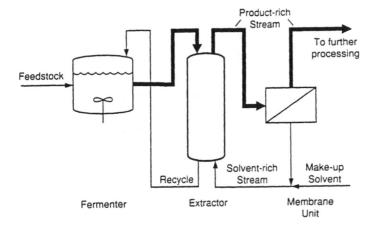

Figure 4   Possible hybrid process for purification of products by extraction and membrane operation (PV, VPe, or RO).

since the permeate does not vaporize; this factor eliminates the need for staged heat input or the design of complex heated modules. VPe has lower energy requirements than PV. Concentration polarization effects are reduced due to the high mass transfer rates of vapors (Jansen et al., 1988). Sander and Soukup (1988b) describe the Lurgi pervaporator designed specifically for use in VPe and discuss VPe process design.

Ethanol/water VPe data for water-selective membranes are presented in Table 17. The separation factors exhibited by VPe membranes can be extremely high, with flux equivalent to PV. Pilot plants using VPe to dehydrate ethanol and other organics are reportedly being built in various parts of the world.

A generalized schematic of a hybrid distillation/VPe process for ethanol recovery is presented in Figure 5. In this process, the ethanol is first concentrated to 70–94 wt % ethanol by fractional distillation. The distillate is partially condensed. The liquid condensate is recycled to the distillation and the vapor is fed to the VPe unit. The retentate is condensed and constitutes the dehydrated ethanol product. The permeate is recycled to the fractional distillation column.

Leeper (1986) reported that Monsanto (St. Louis, MO) was developing VPe membranes/modules for hydration of organic liquids. Monsanto has performed studies on VPe technology but is not presently pursuing commercialization of this technology (Beaver, 1989).

Ethanol/water VPe data for ethanol-selective membranes are presented in Table 18. In theory, ethanol-selective VPe membranes could be used for ethanol dehydration (if very high separation factors can be achieved). The two membranes listed in Table 18 do not appear to exhibit sufficiently high separation factors to be used for ethanol dehydration.

Table 16 Ethanol/Organic Pervaporation Data: Miscellaneous Membranes

Feed mixture and membrane material	Type of data	Ethanol concentration in feed (wt %)	Temperature (°C)	Permeate pressure (kPa)	Separation factor (α_{AB})	Flux (kg/m²·hr)	Comments	Ref.
Ethanol/n-Hexane								
Polyvinyl alcohol (GFT)	J_T, α	10	NR	NR	~1800	~0.022	Ethanol is component A, i.e., ethanol is the component that preferentially permeates.	Fleming (1989a)
Polyvinyl alcohol	J_T, α	10	NR	NR	~900	<0.01	Ethanol is component A.	Schnieder (1987)
Ethanol/n-Heptane								
Cellulose acetate	J_T vs. X_E, α vs. X_E	10–95	23	NR	Comments	1.4–5.6	Ethanol permeates up to $X_E \sim 60$ wt % with low α (~1–4). Heptane permeates at $X_E > 60$ wt %, with low α (~1–3).	Okada (1988)
Ethanol/Toluene								
Polyphenyl-quinoxaline	J_T vs. X_E, α vs. X_E	10–85	70	0.1	2–8	0.2–0.8	Toluene is component A.	Guanwen (1988)

Vapor Permeation: Ethanol Concentration

Ethanol-selective VPe membranes (Table 18) could also be used for concentration of ethanol, but such a process is not practical. Streams containing ethanol at low concentration are generally liquids, which must be vaporized before being fed to the VPe unit. Partial vaporization of the stream will achieve one equilibrium stage of concentration (a substantial degree of ethanol enrichment) before the stream is even fed to the VPe unit. In addition, such an approach will require substantially more energy than that required by PV or reverse osmosis because much more material must be vaporized.

Reverse Osmosis

All RO membranes tested to date are ethanol-rejecting (i.e., preferentially permeate water), with one recent exception (discussed below). RO is being studied for use in ethanol concentration, especially for preconcentrating dilute streams prior to further concentration by other operations. Theoretical studies have been performed on ethanol dehydration using RO, but as yet this application remains impractical. RO may also be useful in hybrid processes with liquid extraction. The ethanol/water RO data published up to 1985 are reviewed in Leeper (1986) and (with the exception of the PEC-1000 membrane) are not reproduced. Data published since 1985 or data omitted from Leeper (1986) are presented in this chapter.

Reverse Osmosis: Ethanol Concentration

Updated ethanol/water RO data are presented in Table 19. Reasonable rejection and high flux are exhibited by the PEC-1000, NS-100, polyacrylamide/polysulfone, and cellulose acetate/styrene-grafted membranes. The flux and rejection of all of these membranes are highest at low ethanol concentrations. Therefore, these membranes may be most useful for partial concentration of dilute ethanol streams. The PEC-1000 and the cellulose acetate/styrene-grafted membranes have been reported to be stable after long-term exposure to ethanol/water solutions.

Tanimura et al. (1990) report data on a RO membrane that exhibits water rejection and preferentially permeates ethanol. This membrane material, n-hexadimethylsilylated poly[1-(trimethylsilyl)-1-propyne] (PTMSP), is the first membrane to exhibit such properties (to the knowledge of the author of this chapter). Such a membrane could potentially be useful for ethanol purification. Tanimura et al. (1990) report RO data for this membrane in McCabe–Thiele–type diagrams (i.e., y vs. x) for ethanol concentrations varying between 22 and 92 wt % (25°C). In addition, flux and water rejection data are reported as a function of feed side operating pressure (3.9–7.8 MPa). Water rejection ranges from 0 to 20 wt %, with flux varying between ~0 and 11 KMH. These membranes exhibit very low ethanol-to-water selectivity. Without major improvements in separation performance, these membranes are not likely to find practical applications for concentrating or dehydrating ethanol. However, this finding is of potential importance. It

Table 17 Ethanol/Water Vapor Permeation Data: Water-Selective Membranes

Membrane material	Type of data	Ethanol concentration in feed (wt %)	Temperature (°C)	Permeate pressure (kPa)
Cellulose acetate	J_T, α	4.0	NR	NR
Cellulose acetate	J_T vs. X_E	0–99	35–70	0.27
	α vs. X_E	28–99	25–70	
	J_T vs. X_E	90–99	35–70	
	J_T vs. T	0–100	25–70	
	J_T vs. PP	90	50	0.07–6.7
	α vs. PP	90	50	
Cellulose acetate (hydrolyzed)	J_T, α	96.0	86	NR
Chitosan	J_T vs. X_E	0–100	40	0.002
	α vs. X_E	44–96		
	Y_E vs. X_E	45–95		
Chitosan/glutaraldehyde (crosslinked)	J_T vs. X_E	0–100	40	0.002
	α vs. X_E	44–95.6		
Chitosan acetate	J_T vs. X_E	77–95.6	40	0.002
	α vs. X_E			
Cellulose	J_T, α	94.0	86	NR
Polyvinyl alcohol	J_T, α	96.7	86	NR
Polysulfonamide	J_T, α	96.0	84	NR
Polysulfonate	J_T, α	99.7	84	NR
Polyamide (FT-30)	J_T, α	96.0	89	NR
Permasep	J_T, α	96.0	86	NR
AKZ0 membrane	J_T vs. X_E	90–99.5	NR	NR
	α vs. X_E			
Lurgi pervaporator	J_W vs. X_E	92–100	100	1.0–3.0

indicates that additional ethanol-selective, water-rejecting membranes may be developed. New materials could exhibit similar flux performance with potentially better separation properties.

The recovery of ethanol from fermentation broth has been studied. Choudhury and Ghosh (1986) studied the cellulose acetate/styrene-grafted membrane

Separation factor (α_{WE})	Flux (kg/m^2·hr)	Comments	Ref.
1000	0.08	Hollow-fiber module.	Jansen (1988)
~200	~0–3	Membrane thickness is 17–19 μm. PP is permeate side pressure.	Suematsu (1988)
	~0.005		
1390	0.21		Jansen (1988)
25–124	0.006–0.176		Uragami (1988)
1500–∞	0.003–0.24		Uragami (1988)
5–2556	0.002–0.116		Uragami (1988)
180	0.04	Membrane thickness: 25 μm.	Jansen (1988)
>2000	0.06		Jansen (1988)
450	0.016	Membrane thickness: 30 μm.	Jansen (1988)
1630	0.006	Membrane thickness: 30 μm.	Jansen (1988)
12.4	3		Jansen (1988)
500	0.02	Membrane thickness: 20 μm.	Jansen (1988)
20–300	0.08–0.13	Hollow-fiber module.	Jansen (1988)
NR	~0–0.4	Membrane material not identified.	Sander (1988b)

using broth containing 8 wt % ethanol. Flux and rejection were compared for aqueous ethanol solutions and for fermentation broth. The performance was consistently lower for fermentation broth. The change in performance was attributed to the complexity of the broth (Choudhury and Ghosh, 1986). Dick and Mavel (1982) compared RO performance of a polyacrylamide/polysulfone membrane

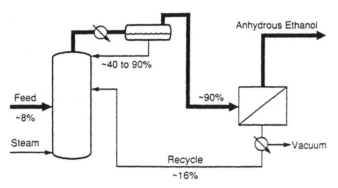

Figure 5   Process for dehydration of ethanol by vapor permeation.

using aqueous ethanol, unclarified broth, and broth clarified by ultrafiltration. The rejection and flux for aqueous solution (2.6 wt % ethanol) were 90% and 27.5 KMH. For unclarified broth (3.15 wt % ethanol), rejection and flux were 86.5% and 15 KMH. For clarified broth (3.0 wt % ethanol), rejection and flux were 86% and 17 KMH. Since the ethanol concentrations were not exactly equivalent, definitive conclusions cannot be drawn. Flux was significantly lower for fermentation broth relative to aqueous ethanol; rejection may be lower. Clarification may enhance flux.

The effect of glucose concentration on ethanol rejection and flux through cellulose acetate RO membranes has been studied by de Pinho et al. (1988). Ethanol rejection decreases as the concentration of glucose in the feedstream increases. Apparently, glucose decreases the solubility of ethanol in water, thereby increasing ethanol solubility in cellulose acetate. This study was performed to evaluate RO for separation of glucose from ethanol (e.g., recovery of unconverted glucose). For such applications, this study is a positive result; the separation of glucose and ethanol can be enhanced by operating at high glucose concentration. For applications in which ethanol is to be recovered from fermentation broth, unconverted glucose concentration should be kept low; then the presence of glucose will not decrease the ethanol rejection (and recovery of ethanol).

A simplified schematic of a hybrid RO/distillation process for concentrating ethanol is presented in Figure 6. This process has been analyzed for feedstreams containing from 2 to 5 wt % ethanol. For feedstreams containing greater than 4.8 wt % ethanol, this process was found to be more expensive and to consume more energy than fractional distillation alone (Mehta, 1982). However, for feedstreams containing ethanol at low concentration (<2–3 wt %), this process is favored over fractional distillation alone (Ohya, 1983a; Leeper and Tsao, 1987; Bitter, 1988). Fermentation of unconcentrated, low-sugar, and low-cellulose feedstocks produces low-proof beer. Ethanol preconcentration of low-proof beer may be of

Table 18 Ethanol/Water Vapor Permeation Data: Ethanol-Selective Membranes

Membrane material	Type of data	Ethanol concentration in feed (wt %)	Temperature (°C)	Permeate pressure (kPa)	Separation factor (α_{EW})	Flux (kg/m²·hr)	Comments	Ref.
Polyvinyl alcohol/ polyacrylonitrile	J_T vs. X_E Y_W vs. X_W	90–98	100	2–15	Moderate	0.5–2.3		Rautenbach (1988)
Polydimethyl- siloxane (silicon rubber)	J_T, α J_T vs. T	44	25 25–70	0.002	5.6 0.8–7.2	0.08–0.12		Uragami (1988)

Table 19   Ethanol/Water Reverse Osmosis Data: Ethanol-Rejecting Membranes

Membrane material	Type of data	Ethanol concentration in feed (wt %)	Temperature (°C)	Feed pressure (MPa)
Polyfurane/polycyanurate (PEC-1000)	J_T vs. X_E R_E vs. X_E	2–50	~20	2.0–7.8
Polyfurane/polycyanurate (PEC-1000)	R_E vs. X_E R_E vs. FP	2.6–18.2	—	6.0 6.0–10
Cellulose acetate	J_T vs. X_E J_T vs. FP R_E vs. X_W R_E vs. FP	4–16	30	5.5–9.6
Cellulose acetate	J_T vs. X_E R_E vs. X_E	0.005–0.02	24 ± 1	1.72
Celloluse acetate	R_E	1.0	25	4.9
Cellulose acetate/styrene-grafted	J_T vs. X_E J_T vs. FP R_E vs. X_E R_E vs. FP J_W , J_E vs. X_E	4–16 8 4–16 8 4–16	28 ± 2	8.3
Cellulose acetate/styrene-grafted	R_E vs. X_E T_E vs. FP J_T vs. X_E J_T vs. FP	4–16	28 ± 2	5.5–9.6
Cellulose acetate/styrene-grafted	J_T vs. X_E R_E vs. X_E J_T, R_E vs. X_S J_T, R_E vs. RD J_T, R_E vs. pH J_T, R_E vs. T	4–16 8 8	30 20–50	8.3
Polyacrylamide/ polysulfone	J_T vs. X_E R_E vs. X_E	1–9	25	6.0
Polyacrylamide/ polysulfone	J_T, R_E	2.5	25	6.0
Polyethylenimine/*m*-tolylene-2,4-disocyanate on polysulfone	J_T, R_E vs. P J_T, R_E vs. T	4.6	25–30	2.0–7.8
Polyvinyl alcohol/polystyrene sulfonic acid	R_E	0.02	25	7.88
Polybenzimidazolone	R_E	1.0	25	4.9

Ethanol rejection (%)	Flux (kg/ m²·hr)	Comments	Ref.
20–92	0.1–15.6	Data combined from both references.	Ohya (1981) Ohya (1982)
50–99	NR	Overall rejection for concentration from 0.5 wt % to 10 wt % ethanol is 85%. FP is feed side pressure.	Ohya (1983a)
10–35	7–19	FP is feed side pressure.	Choudhury (1988)
15	19–32	Composite data for two different membranes of same type.	Matsuura (1977)
32	NR		Murakami (1981)
	1.6–2.1	FP is feed side pressure.	Choudhury
	0.8–2.4		(1985)
60–90			
80–90			
70–90	0.4–1.2	These data are for aqueous solutions of ethanol; performance is lower for actual fermentation broth. This membrane reported to be stable after prolonged exposure to 16 wt % ethanol. FP is feed side pressure.	Choudhury (1986)
80–90	~2	X_S is styrene content of membrane. RD is radiation dose of membrane. pH of feed varied between 4 to 10.	Choudhury (1988)
80–99	2.5–11	Composite data for several different membranes of same type.	Dick (1982)
60–65	25–29	Different membrane from data given above.	Dick (1982)
~60	8–40	Essentially the NS-100 membrane.	Huang (1984)
40	NR		Koyama (1982)
61	NR		Murakami (1981)

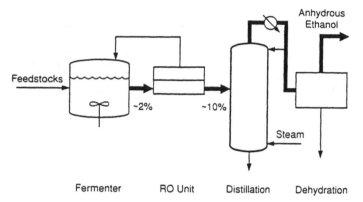

Figure 6   Hybrid process for recovery of fermentation product by reverse osmosis and distillation.

practical importance. No commercial installations of RO for ethanol preconcentration are described in the literature. However, this application of RO has been analyzed.

The energy requirements of ethanol concentration by RO have been estimated by Ohya (1983a). The analysis was based on the PEC-1000 membrane, operated at 6.0 MPa. The estimates were made for concentration of three feedstreams to 15 wt % ethanol. For starting concentrations of 0.5, 1.0, and 2.0 wt % ethanol, the energy requirements were 4.4, 2.2, and 1.1 GJ/m^3 (ethanol basis), respectively.

Leeper and Tsao (1987) analyzed the use of RO preconcentration in the production of ethanol from low-cellulose lignocellulose feedstocks. The feedstocks was wheat straw and the final beer concentration was 1.9 wt % ethanol. Two ethanol purification processes were compared: (a) fractional distillation (to 92.4 wt %) plus azeotropic distillation (to anhydrous) and (b) RO (to 8 wt %), fractional distillation (to 92.4 wt %) and azeotropic distillation (to anhydrous). The RO operation was based on the PEC-1000 membrane operated at 5.88 MPa. The operating costs (including capital) of the RO-based process were 3% lower than those of the distillation process. Incremental energy requirements of the RO operation were 0.3 GJ/m^3, which is in line with the estimates of Ohya (1983a). Total energy requirements (feedstock preparation through ethanol recovery) for the process using RO preconcentration were 16 GJ/m^3 of anhydrous ethanol. The conventional process (Figure 1) requires about 35 GJ/m^3.

Bitter (1988) also analyzed RO preconcentration of low-proof beer (4 wt % ethanol). A multistage flash distillation (MSFD)/azeotropic distillation (AD) process was compared to a hybrid RO/MSDF/AD process. In the RO-based process, the ethanol was preconcentrated to 10 wt % prior to MSDF. The RO operation was based on the PEC-1000 membrane operated at 7.0 MPa. The energy cost of ethanol purification by MSDF/AD process was $365/m³ anhydrous ethanol ($1.38/gallon). The total production cost was about $540/m³ ($2.04/gallon). The hybrid RO/MSDF/AD process had an energy cost of about $137/m³ ($0.52/gallon) and a total production cost of about $375/m³ ($1.42/gallon). The incremental energy requirement of RO was 0.64 GJ/m³. The difference in energy requirements between the analysis of Bitter (1988) and Leeper and Tsao (1987) is generally accounted for by the different RO operating pressures and concentration ranges. Bitter (1988) estimates that RO preconcentration reduces total production costs about 30%.

For low-sugar feedstocks, sugar concentration prior to fermentation should also be considered. These two approaches—(a) feedstock preconcentration to produce ethanol at higher concentration and (b) unconcentrated feedstock fermentation followed by ethanol preconcentration by RO—have not been directly compared. However, Zacchi and Axelsson (1989) analyzed feedstock preconcentration. The economics of ethanol production using feedstocks containing 1.5–5 wt % glucose were compared for (a) feedstock preconcentration by RO, (b) feedstock preconcentration by six-stage evaporation, and (c) no feedstock preconcentration. Feedstock preconcentration was varied up to 16 wt % glucose. In all cases, ethanol was recovered by conventional distillation. Evaporative preconcentration had the highest costs and energy requirements. RO preconcentration had the lowest costs and energy requirements. For instance, continuous fermentation of a feedstock containing 1.5 wt % glucose had a production cost of about $610/m³ ($2.30/gallon) and an energy requirement of 25.7 GJ/m³ (92,300 Btu/gallon) (anhydrous ethanol basis). The RO preconcentration (from 1.5 to 8.0 wt % glucose)/continuous fermentation process had a production cost of $425/m³ ($1.61/gallon).

The continuous membrane column concept has been studied and analyzed for use in ethanol concentration by RO. This concept was first described for RO by Loeb and Bloch (1973). Lee et al. (1983, 1984) call this RO operation continuous countercurrent RO (CCRO). Bitter (1988) calls it countercurrent reflux RO (CCRRO). In CCRO, a fraction of the retentate stream is recycled on the permeate side of the membrane to increase the ethanol concentration in that stream. Since the ethanol concentration difference across the membrane is reduced, the osmotic pressure difference is also reduced and the net driving force is increased.

Ethanol concentration by CCRO was first studied, and demonstrated, by Lee et al. (1983, 1984). In order for the CCRO concept to work, the membrane must have low resistance to diffusion of ethanol from the permeate stream to the permeate side of the membrane layer. The ethanol concentration at the permeate side of the membrane must be representative of the bulk concentration of ethanol in the permeate stream. The net reduction in osmotic pressure difference across the membrane is determined by the ethanol concentration at the permeate side of the membrane, not by the bulk concentration of ethanol in the permeate stream. Polyamide membranes allowed the most significant CCRO effect. The economics of the CCRO process are potentially favorable (Lee et al., 1983).

Bitter (1988) also analyzed the CCRO concept for preconcentrating ethanol from 4 to 10 wt %. A hybrid CCRO/MSDF/PV process for production of anhydrous ethanol had a total production cost of about $256/m^3 ($0.97/gallon) and an energy cost of $130/m^3 ($0.49/gallon). An equivalent MSDF/AD process had costs of $523/m^3 and $354/m^3, respectively. Bitter (1988) notes that CCRO is not being used commercially at any installations in the world.

Reverse Osmosis: Ethanol Dehydration

The use of RO for dehydration of ethanol has been proposed by Gregor and Jeffries (1979) and Mehta (1982). However, suitable water-rejecting RO membranes have not yet been developed. The n-hexyldimethylsilylated poly[1-(trimethylsilyl)-1-propyne] membrane studied by Tanimura et al. (1990) does exhibit water rejection. However, the rejections reported to date are very low (approaching 1.0 for higher ethanol concentrations). It has not been studied for ethanol concentrations above 80 wt %. As presently developed, this membrane does not appear to be a viable candidate for dehydrating ethanol. However, the identification of one water-rejecting RO membrane indicates that other materials with potentially higher performance may be developed.

Mehta (1982) performed a preliminary economic analysis of ethanol dehydration using a hypothetical RO membrane. In the Mehta (1982) study, only the costs of ethanol dehydration were compared, as RO costs vs. azeotropic distillation costs. On this basis, Mehta reported that RO looked promising—if an appropriate membrane could be developed.

Leeper and Tsao (1987) performed a more detailed analysis of RO for ethanol dehydration in which the costs of a complete plant were considered. Capital and production costs of dehydration by RO were not reduced relative to azeotropic distillation. The reduced costs of the RO operation were offset by increased costs for fractional distillation. Fractional distillation costs were increased due to recycle of the retentate stream from the RO operation. The overall energy savings derived from use of RO dehydration were about 0.8 GJ/m^3 (anhydrous ethanol basis). Leeper and Tsao (1987) conclude that there is little incentive for development of an RO membrane for dehydration of ethanol.

Reverse Osmosis: Ethanol/Organic Separations

Hybrid RO/extraction processes can also be envisioned. RO could be used to separate ethanol from solvent or extraction could be performed using RO-concentrated streams as the feed. Very little ethanol/organic RO data have been published. Leeper (1986) presents data for cellulose acetate and polyethylene for separation of ethanol and xylene and ethanol and n-heptane.

Membrane Extraction

Membrane-assisted extraction (sometimes called perstraction) can overcome or reduce the solvent toxicity problem associated with the use of extraction to recover products during continuous fermentation. In theory, membrane extraction maintains separation of the fermentation broth from the solvent while allowing extraction to occur across the membrane (Figure 7). As a result, solvent does not mix with the fermentation broth. Therefore, membrane extraction can permit the use of solvents that are toxic to microorganisms. The advantages of membrane extraction over conventional extraction for the recovery of fermentation products are discussed by Jeon and Lee (1989). The design of hollow-fiber solvent extractors is discussed by Prasad and Sirkar (1990).

Matsumura and Markl (1984) studied perstraction for removal of ethanol from fermentation broth during continuous fermentation. The following membrane materials were screened: cuprous ammonium rayon, ethylenevinyl alcohol, polyvinyl alcohol, cellulose diacetate, polyacrylonitrile, and polysulfone. The hollow-fiber cuprous ammonium membrane was selected. Of the 25 solvents investigated, 2-ethyl-1-butanol, *sec*-octanol, and tri-*N*-butyl phosphate were selected for further study. Using a simulated feed (7.8 g ethanol/liter), ethanol extraction was demonstrated at 30°C. Extraction was also demonstrated using fermentation broth. In fermentations using glucose feed concentrations of 340 g/liter and 470 g/liter, ethanol inhibition was eliminated by perstraction.

This concept has also been studied by Kang et al. (1990b). Yeast cells were immobilized in the shell side of a hollow-fiber module (Celgard X-20). Solvent (oleyl alcohol) was circulated through the fiber lumen. The effects of solvent flow rate, substrate concentration and flow rate, and cell concentration were studied.

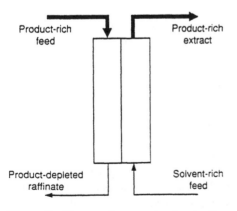

Figure 7   The membrane-assisted extraction concept (perstraction).

Glucose conversion was increased in this mode of operation relative to batch fermentation.

Qin et al. (1990) studied the use of gas membranes to extract ethanol and other products from fermentation broth. In this concept, a gas is suspended in a hydrophobic microporous hollow-fiber membrane. Acids and organics in the fermentation broth volatilize into the gas phase and diffuse across the membrane. The permeating components are then dissolved in a sodium hydroxide solution on the permeate side of the membrane. Candidate gases included sulfur dioxide, bromine, hydrogen sulfide, and hydrogen cyanide. Ethanol was extracted from water using this approach.

ACETONE, BUTANOL, AND ISOPROPANOL RECOVERY

The acetone-butanol-ethanol (ABE) and isopropanol-butanol-ethanol (IBE) fermentations are also of industrial interest. Recent reviews of the ABE fermentation include Linden et al. (1985), Jones and Woods (1986), McNeil and Kristiansen (1986), Ennis et al. (1986), and Awang et al. (1988). The conventional ABE fermentation and recovery process is described by Beesch (1952) and Walton and Martin (1979). The conventional distillation technology for separation of acetone, butanol, and ethanol [also briefly described by Leeper (1986)] requires about 28–42 GJ/m^3 of anhydrous products (100,000–150,000 Btu/gallon). The IBE fermentation is reviewed by Haggstrom (1985).

Membrane applications in the recovery of acetone and butanol were reviewed by Leeper (1986). Due to the limited data, acetone and butanol data reported in Leeper (1986) are also reported in this chapter. Isopropanol recovery was not discussed in Leeper (1986).

Pervaporation

Pervaporation is being studied for dehydration of acetone, butanol, and isopropanol. PV is also being studied for the preconcentration of these three chemicals from aqueous solutions.

Pervaporation: Product Dehydration (Water-Selective Membranes)

Commercially available PV membranes for dehydration of organic liquids include the GFT polyvinyl alcohol (PVA) membrane (Fleming, 1989a, 1989b) and the tubular, Kalsep BP membrane (Pearce, 1988).

Acetone/water PV data for water-selective membranes are presented in Table 20. The GFT PVA membrane exhibits suitable flux and high separation factors. The carboxymethylcellulose/polyacrylic acid (sodium counterion) membrane may be useful, but additional data are required.

Table 20 Acetone/Water Pervaporation Data: Water-Selective Membranes

Membrane material	Type of data	Acetone concentration in feed (wt %)	Temperature (°C)	Permeate pressure (kPa)	Separation factor (α_{WA})	Flux (kg/m²·hr)	Comments	Ref.
Polyvinyl alcohol (GFT)	Y_A vs. X_A	0–100	30	NR	Moderate to high	NR	α extremely high for $X_A > 70$ wt %.	Fleming (1989b)
Carboxymethyl cellulose/ polyacrylic acid (Na$^+$ counterion)	PR, α	89	25	NR	5500	~0.4	PR is permeation rate (g·mm/ m²·hr). Flux estimated from approximate membrane thickness (1 mm).	Reineke (1987)
Polyvinyl alcohol	Y_A vs. X_A	0–1.0	60	2.0	Low to moderate	NR	Water-selective for $X_A > 25$ wt %.	Wesslein (1990)

Butanol/water PV data for water-selective membranes are presented in Table 21. Cellulose acetate exhibits high selectivity and reasonable flux for PV. Again, the carboxymethylcellulose/polyacrylic acid (sodium counterion) membrane may be useful, but additional data are required.

Isopropanol/water PV data for water-selective membranes are presented in Tables 22 (cellulose membranes) and 23 (miscellaneous membranes). The GFT PVA membrane is a proven commercial membrane for isopropanol dehydration. The Nafion membrane exhibits good performance. The performance of ion exchange (e.g., Nafion) membranes is highly dependent on the counterion incorporated into the membrane (Cabasso and Liu, 1985). The performance of PV membranes is also highly dependent on postformation treatment (Nagy et al., 1983). Hydrated cellulose exhibits very different performance if treated with isopropanol or water (Table 22). The isopropanol-treated membrane exhibits high separation factors and flux, whereas the water-treated membrane has low separation factors. Additional interesting PV membranes include polyphenylquinoxaline, the Kalsep BP membrane, and the Texaco membrane. It could be interesting to see additional data on the carboxymethylcellulose/polyacrylic acid (sodium counterion) membrane.

PV plants for dehydrating acetone, n-butanol, and isopropanol are presently operating (Asada, 1988; Fleming, 1989b). For example, aqueous acetone (96.2 wt %) is dehydrated to 99.99 wt % using PV and aqueous n-butanol (91.6–98.6 wt %) is dehydrated to 99.92–99.96 wt % using the GFT PVA membrane (Fleming, 1989b). A plant in Japan uses the GFT PVA membrane to produce 4.5 m³/day of anhydrous acetone; the feed contains 99.5 wt % acetone and the final concentration is 99.7 wt % (Asada, 1988).

The isopropanol-water azeotrope occurs at 88 wt % isopropanol at atmospheric pressure (Perry, 1963; Horsley, 1973). A simplified schematic of a hybrid distillation/PV process for dehydration of isopropanol is shown in Figure 8; this process is being used commercially (Fleming, 1989b). PV can also be used as the isopropanol dehydration step. A plant in Japan is using PV to dehydrate (to 99.7 wt %) an isopropanol feed containing 87 wt %; the capacity of this plant is 15.2 m³/day (Asada, 1988).

Sander and Soukup (1988b) estimate that the cost of dehydrating isopropanol by PV is $0.26/kg; extractive distillation is estimated to cost $0.34/kg.

Pervaporation: Product Concentration (Organic-Selective Membranes)

The major application of organic-selective PV membranes involves solvent removal from continuous ABE and IBE fermentations; this approach reduces feedback inhibition and can thereby increase productivity.

Acetone/water PV data for acetone-selective membranes are presented in Table 24. The polymethoxysilane membrane, which is commercially available

Table 21 Butanol/Water Pervaporation Data: Water-Selective Membranes

Membrane material	Type of data	Butanol concentration in feed (wt %)	Temperature (°C)	Permeate pressure (kPa)	Separation factor (α_{WB})	Flux (kg/m²·hr)	Comments	Ref.
Carboxymethyl-cellulose/polyacrylic acid (Na⁺ counterion)	PR, α	89	25	NR	730	~0.4	PR is permeation rate (g·mm/m²·hr). Flux estimated from approximate membrane thickness (1.0 mm).	Reineke (1987)
Cellophane	J_T vs. X_B α vs. X_B	10–24	60	2.7	10–15	1.0–2.0	Flux and selectivity estimated from data presented graphically.	Nagy (1980)
Cellulose acetate	J_T vs. X_B α vs. X_B	60–95	60	4.0	30–100	0.05–0.15	Flux and selectivity estimated from data presented graphically.	Stelmaszek (1982)
Cellulose acetate	J_T, α	85	60	2.7	45	1.0		Nagy (1980)

Table 22 Isopropanol (IPA)/Water Pervaporation Data: Cellulose Membranes (Water-Selective)

Membrane material	Type of data	Isopropanol concentration in feed (wt %)	Temperature (°C)	Permeate pressure (kPa)
Cellulose acetate	J_T vs. X_P	45–100	25	0.4
	α vs. X_P	45–91		
	Y_P vs. X_P			
	J_W, J_E vs. X_P			
	Y_P, Z_P vs. Θ			
Cellulose acetate	Y_P vs. X_P	5–85	25	1.3–5.3
Cellophane	J_T vs. X_W	27–93	60–70	NR
	Y_W vs. X_W			
Cellulose triacetate	J_T vs. X_W	16–30	20	0.01
Cellulose-hydrated (IPA-treated)	J_T vs. X_W	22–95	60	2.0
	Y_W vs. X_W	24–96		
	J_T vs. PP			1.0–50
	Y_W vs. PP			1.0–50
Cellulose-hydrated (water-treated)	J_T vs. X_W	22–95	60	2.0
	Y_W vs. X_W	22–88		
Carboxymethylcellulose/ polyacrylic acid (Na$^+$ counterion)	PR, α	89	25	NR

(GFT), exhibits moderate separation factors at low acetone concentrations; this membrane may be useful for concentrating acetone. Polypropylene exhibits high flux but low separation factors.

Butanol/water PV data for butanol-selective membranes are presented in Table 25. At low butanol concentrations, silicone rubber exhibits high separation factors but low flux. The polyether-block-amide membrane has high PV flux but low to moderate separation factors.

Isopropanol/water PV data for isopropanol-selective membranes are presented in Table 26. Silicone rubber exhibits low to moderate separation factors at low isopropanol concentrations.

Various studies have demonstrated that productivity of ABE and IBE fermentations can be increased by using PV to continuously remove solvents. These studies indicate that performance is lower when the feed is actual fermentation broth (as compared to aqueous solutions). The use of silicone rubber PV membranes in the ABE fermentation was studied by Sodeck et al. (1987) and Larrayoz

Separation factor (α_{WP})	Flux (kg/m^2·hr)	Comments	Ref.
1–12	0.04–0.36	Z_p is propanol concentration in retentate. Θ is stage cut. α_{WP} is low for $X_p < 80$ wt % and moderate for $X_p > 80$ wt %.	Seok (1987)
Moderate	NAR	NAR: Not adequately reported.	Shiyao (1988)
Moderate to low	0.1–5	Low α at high X_p.	Carter (1964)
15–55	0.1–0.19		Changlou (1989)
High	0.04–1.4	PP is permeate pressure. Flux is essentially zero for PP > 40 kPa.	Nagy (1983)
	~0.22–5.8		
High	0.04–1.4	PP is permeate pressure.	Nagy (1983)
>10,000	~0.16	PR is permeation rate (g·mm/m^2·hr). Flux estimated from approximate membrane thickness (1 μm) and PR.	Reineke (1987)

and Puigjner (1987). Groot et al. (1984a, 1984b, 1988) and Groot and Luyben (1987) studied silicone rubber PV membranes for continuous product recovery during batch and continuous IBE fermentations. PV removal of solvents allows complete glucose utilization (or shortens fermentation time) for batch operation or allows higher feedstock concentration for continuous fermentations (for any given butanol concentration). Productivity increases of 33–70% are reported.

Schoutens and Groot (1985) performed a preliminary economic analysis of a process using PV to remove products during a continuous IBE fermentation. The plant capacity was 45 kg IBE (99.5 wt %)/year. The design was based on laboratory and literature data for immobilized *Clostridium beyerinckii* and silicone rubber PV membranes. Purification was completed by distillation and PV. Using an overall products-to-water separation factor of 50, the authors conclude that this process could be economically feasible.

Membranes that exhibit sufficiently high separation factors for all of these products are not presently developed. Higher flux may be necessary.

Table 23 Isopropanol (IPA)/Water Pervaporation Data: Miscellaneous Membranes (Water-Selective)

Membrane material	Type of data	Isopropanol concentration in feed (wt %)	Temperature (°C)	Permeate pressure (kPa)
Polyvinyl alcohol (GFT)	J_T vs. X_P, PP Y_p vs. X_p, PP	0–100	60	3–13
Polyvinyl alcohol (GFT)	J_T vs. X_P Y_p vs. X_p	74–98.5	70	NR
Polyvinyl alcohol	Y_p vs. X_p	1–99	60	3
Polyvinyl alcohol/ polyacrylonitrile	Y_W vs. X_W	10–90	80	2
Nafion (DuPont) (Cs$^+$ counterion)	J_T vs. X_W α vs. X_W J_T vs. X_W J_T, α vs. CI Y_W vs. X_W	0–100 0–100 88–99 — 8–90	30	<0.1
Nafion (DuPont) (K$^+$ counterion)	J_T vs. X_W Y_W vs. X_W	50–95 53–98	40	<0.1
Sulfonated polyethylene (Cs$^+$ counterion)	J_T vs. T	88–90	26	>0.1
Polyacrylic acid/ polyacrylonitrile on cellulose acetate	J_T, α	14.6	40	0.01
Polyvinyl acetate/poly- sulfone (composite)	J_T vs. X_W α vs. X_W	~0–1.5	40	0.01
Polysulfone/cellulose acetate (composite)	J_T vs. X_W α vs X_W	5.6–8.7	40	0.01
Polyphenylquinoxaline (ether linkages)	J_T vs. X_p α vs. X_p	8.3–85	70	0.1
Polybenzimidazole	J_T vs. X_W Y_W vs. X_W	98.4–99.7	30	NR
Polyethylene-co- styrenesulfonate (K$^+$ counterion)	J_T vs. X_W Y_W vs. X_W	59–98 50–90	40	<0.1
Polytetramethylene glycol-block-nylon-12	Y_p vs. X_p α vs. X_p	1–5	NR	NR
Unidentified membrane (Kalsep BP)	J_W vs. X_W Y_W vs. X_W J_T vs. T	70–99 90–99 95	70 50–80	NR
Unidentified membrane (Texaco)	J_T vs. X_p Y_p vs. X_p J_T, α vs. t	74–98.5	70	NR

Separation factor (α_{WP})	Flux (kg/m²·hr)	Comments	Ref.
Moderate to high	~0–0.4	Water-selective for $X_p > 20$ wt %. J_T and α highest at low permeate side pressure. Flux approaches zero as X_W approaches zero. α high at low X_p. GFT is manufacturer. PP is permeate side pressure.	Wesslein (1988, 1990)
High	~0–0.6	GFT is manufacturer.	Bartels (1988)
Moderate to very high	NR		Hauser (1989)
Moderate to high	NR		Rautenbach (1988)
1–36	0.1–1.8	Nafion is perfluorosulfonic acid on polytetrafluoroethylene.	Cabasso (1985a; 1986)
	0.03–0.5	Nafion 811 hollow fibers (DuPont), with 90 to 103 μm wall thickness. Cs is counterion; study includes Li, Na, K, Ca, Al.	
Moderate to high	~0–1		Kujawski (1988)
~20,000	0.1–0.5	Additional cations studied (K, Na, Li, H). H^+ has highest J but lowest α. Na^+ has highest α but lower J.	Cabasso (1985b)
43.2	0.13		Changlou (1989)
1118–28629	0.004–0.006		Changlou (1989)
129–326	0.007–0.016		Changlou (1989)
7–130	0.15–0.4		Guanwen (1988)
High	0.012–0.017	Membrane thickness (~53 μm).	Sferrazza (1988)
High	~0–0.3	Highest α at low X_p.	Kujawski (1988)
1.3–2.0	NAR	NAR: Not adequately reported. α depends on composition; this membrane, designated 5533, is water-selective at $X_p = 1$ wt % and IPA selective at $X_p = 5$ wt %.	Morin (1988)
High	~0–2.1	Flux is lowest at low X_p.	Pearce (1988)
High	~0–2	α highest for $X_p > 85$ wt %. t is time; data for 45 hr at $X_p = 82$ wt % and 80°C.	Bartels (1988)

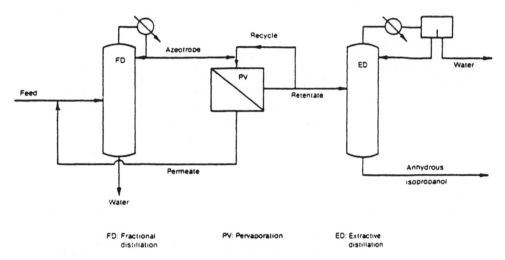

Figure 8 Hybrid process for dehydration of isopropanol. (From Fleming, 1989a.)

Table 24 Acetone/Water Pervaporation Data: Acetone-Selective Membranes

Membrane material	Type of data	Acetone concentration in feed (wt %)	Temperature (°C)	Permeate pressure (kPa)
Polymethoxysilane (GFT)	Y_A vs. X_A	0–100	30	NR
Polypropylene	J_T, α vs. T	45	84–130	6.5
Silicone rubber (PDMS)	Y_A vs. X_A	1–10	30	~1.0
Silicone rubber (PDMS)	α	1.0	NR	NR
Silicone rubber (PDMS)	J_A vs. X_A J_T vs. X_A α vs. X_A	0.5–0.6	37	NR
Polyolefin/PDMS	Y_A vs. X_A	0–1	30	~1.0
PTMSP	α, J_T Y_A vs. X_A	10 10–80	30	.01
Oleyl alcohol/polyethylene liquid membrane	J_T vs. X_A α vs. X_A Y_A vs. X_A	1–16	30	0.13

Vapor Permeation: Products Dehydration

Very little VPe data are available for evaluation of acetone, butanol, and iso-propanol recovery applications (Table 27). The PVA/polyacrylonitrile membrane exhibits moderate to high separation factors for water and low to high PV flux in the isopropanol/water separation.

Pilot scale studies were performed on isopropanol dehydration by VPe (Sander and Soukup, 1988b). The pilot plant, VPe module, and VPe process design are described. Results of an economic evaluation indicate that VPe has lower production costs ($0.16/kg) for isopropanol dehydration than either extractive distillation ($0.34/kg) or PV ($0.26/kg).

Reverse Osmosis: Products Concentration

RO has been studied for concentration of acetone, butanol, and isopropanol. All membranes studied to date reject the organic components of an aqueous solution.

Acetone/water and butanol/water RO data are presented in Tables 28 and 29, respectively. Additional data for both systems would be useful. The FT-30 polyamide membrane may be useful for product recovery in ABE and IBE fermentations.

Separation factor (α_{AW})	Flux (kg/m²·kr)	Comments	Ref.
Low to moderate	NR	α_{AW} moderate at low X_A, low at high X_A.	Fleming (1989b)
~3	0.1–1.2	Feed pressure was 0.45 MPa.	Featherstone (1971)
Moderate	NR	PDMS is polydimethylsiloxane (silicone rubber). Membrane thickness: 3.5 μm.	Blume (1990)
170	NR	PDMS is silicone rubber.	Watson (1990)
	0.005–0.03	PDMS is silicone rubber.	Larryayoz (1987)
30–40			
High	NR	PDMS is silicone rubber. Membrane thickness: 3.5μm.	Blume (1990)
76	2.2	PTMSP is poly[1-(trimethylsilyl)-1-propyne]. Membrane thickness is ~30 μm.	Masuda (1990)
40–160	0.03–0.38	Highest selectivity at low X_A. Membrane thickness is ~25 μm.	Matsumura (1987)

Table 25   Butanol/Water Pervaporation Data: Butanol-Selective Membranes

Membrane material	Type of data	Butanol concentration in feed (wt %)	Temperature (°C)	Permeate pressure (kPa)
Polyether-block-amides	J_T vs. X_B	0.5–6	50	<0.2
	J_B vs. X_B			
Silicone rubber (PDMS)	J_T vs. X_B	0.4–1.7	37	NR
	α vs. X_B			
Silicone rubber	J_T, α	0.55	37	NR
Silicone rubber	F vs. X_B	0–8	30	NR
	α vs. X_B			
Silicone rubber	J_T vs. X_B	0–10	25	NR
	J_T vs. X_B	85–100		
	α vs. X_B	5–10		
	α vs. X_B	85–100		
	Y_B vs. X_B	5–10		
	Y_B vs. X_B	85–100		
Silicone rubber	J_B vs. X_B	1.4–1.8	37	NR
	J_T vs. X_B			
	α vs. X_B			
Silicone rubber	α vs. X_B	0–1.0	NR	NR
Oleyl alcohol/polyethylene liquid membrane	J_T, α	1.0	30	0.13
	Y vs. X_B	1–4		
Di-n-butyl phthalate/poly-ethylene liquid membrane	J_T, α	1.0	30	0.13
Tricresyl phosphate/poly-ethylene liquid membrane	J_T, α	1.0	30	0.13

Garcia et al. (1986) studied the FilmTec FT-30 membrane for butanol removal during continuous ABE fermentation. The proposed process involves clarification of the broth by ultrafiltration (UF). The UF retentate is recycled to the fermenter, increasing cell concentration and productivity. The UF permeate, which contains acetone and butanol, is treated by RO to recover butanol in the RO retentate. The RO permeate is also recycled to the fermenter. For fermentation broth, flux was about 30% lower than with aqueous solutions. Butanol removal by RO reduced feedback inhibition, allowed use of higher feedstock concentrations, and increased productivity. Overall, the ABE products were concentrated about threefold at a recovery of 70%.

Separation factor (α_{BW})	Flux (kg/m^2·hr)	Comments	Ref.
Low to moderate	0.24–1.75	Low α at low X_B.	Bengston (1988)
45–57	0.05–0.2	Silicone rubber is polydimethylsiloxane (PDMS). Studies included pervaporation of butanol from fermentation broth; performance is generally lower at any given X_B.	Groot (1984a)
11	NR	Feed to pervaporation was fermentation broth containing butanol at 5.5 g/liter.	Groot (1987)
45–65	0.006–0.035	F is permeate flowrate; flux calculated from F and effective membrane area.	Groot (1988)
	~0.01	Flux at X_B = 5 wt %.	Tanigaki (1987)
	~0.06	Flux at X_B = 85 wt %.	
~80		α at X_B = 5 wt %.	
~1			
46–58	0.01–0.03		Larrayoz (1987)
8–500	NR		Watson (1990)
180	0.08	Unstable for X_B > 5 wt %. For X_B < 5 wt %, stable for 100 hr.	Matsumura (1987)
90	0.112		Matsumura (1987)
105	0.055		Matsumura (1987)

Isopropanol/water RO data are presented in Table 30. The PEC-1000 and NS-100 membranes exhibit phenomenal flux and rejection with feeds containing 1.0 wt % isopropanol. Additional study of these membranes is warranted.

Membrane Extraction

The use of membrane-assisted extraction (Figure 7) for recovery of butanol from ABE fermentation broth was investigated by Jeon and Lee (1989). They identify several barriers to the extended use of extraction for the recovery of fermentation products. The solvents with the highest distribution coefficients are frequently also

Table 26   Isopropanol (IPA)/Water Pervaporation Data: IPA-Selective Membranes

Membrane material	Type of data	Isopropanol concentration in feed (wt %)	Temperature (°C)	Permeate pressure (kPa)
PTMSP	J vs. X_P α vs. X_P Y_P vs. X_P	10–85	30	>0.1
Silicone rubber	J_T vs. X_P α vs. X_P J_T vs. T J_T vs. PP P vs. V	9–100 8–79	25	~0.67
Silicone rubber	α vs. X_P Y_P vs. X_P Y_P, Z_P vs. Θ	27–100	25	0.33
Silicone rubber	J_T vs. X_P α vs. X_P Y_P vs. X_P	0–100 20–90	25	NAR
Silicone rubber (PDMS)	Y_P vs. X_P	10–90	30	~1.0
Polyolefin/PDMS	Y_P vs. X_P	10–90	30	~1.0
Polytetramethylene glycol-block-nylon-12	α vs. X_P Y_P vs. X_P	1–5	NR	NR

toxic to the fermentation microorganisms. Since the solvent in conventional extraction is mixed with the feedstream (i.e., the fermentation broth), a fraction of the solvent ends up in the fermentation broth—which leads to inhibition of or damage to the cells. The solvent may also need to be sterilized, which can lead to undesirable chemical changes. Cells and enzymes that contact the water–solvent interface can be inactivated. Cell adsorption at the water–solvent interface can adversely affect the equilibrium distribution coefficients of fermentation products. In order to minimize product recovery costs, the solvent phase must have a density substantially different from that of the aqueous phase. Membrane extraction may provide an approach that expands the use of extraction in the recovery of fermentation products.

Separation factor (α_{PW})	Flux (kg/m²·hr)	Comments	Ref.
1–24	NAR	PTMSP is poly[1-(trimethylsilyl)-1-propyene]. NAR is not adequately reported.	Ishihara (1983)
9–22	0.03–0.11	Silicone rubber is polydimethylsiloxane (PDMS). PP is permeate side pressure. P is performance (J_T and α). V is feed side velocity; varied from 0–50 cm/sec and P found to be relatively independent of flow rate.	Kimura (1983)
0.5–12	NR	Z_p is IPA concentration of retentate. Θ is stage cut. α moderate for $X_p <$ 48 wt %, low for 48% $< X_p <$ 93%. Silicone rubber is water-selective at low α for $X_p >$ 93 wt %.	Seok (1987)
~1–8	~0–0.15	NAR: Not adequately reported.	Tanigaki (1987)
Moderate to low	NR	PDMS is silicone rubber. Membrane thickness: 3.5μm. Becomes water-selective for $X_p >$ 80 wt %.	Blume (1990)
Low	NR	PDMS is silicone rubber. Membrane thickness: 3.5 μm. Becomes water-selective for $X_p >$ 80 wt %.	Blume (1990)
2.3–4.4	NAR	NAR: Not adequately reported. α depends on composition; these two IPA-selective membranes are designated 3533 and 4033. These data are combined data for two different membranes of the same type.	Morin (1988)

In the studies by Jeon and Lee (1989), glucose (at 20–40 g/liter) was converted to ABE. The final concentration of ABE was ~12 g/liter (6–8 g butanol/liter). Broth was pumped through silicone rubber tubing placed in a stirred beaker, containing oleyl alcohol as the solvent. Volumetric productivity was increased dramatically as compared to batch operation. Over several runs, cells eventually became inactive after 168–385 hr of continuous operation. Apparently inhibitors are not removed by extraction. This observation and its implications are discussed in detail by Jeon and Lee (1989).

Shukla et al. (1989) studied the use of membrane extraction for the recovery of ABE. The solvent was 2-ethyl-1-hexanol. The volumetric productivity of the extractive fermentation was 40% higher (0.92 g ABE/liter·hr) as compared to

Table 27 Isopropanol (IPA)/Water Vapor Permeation Data: Water Selective Membranes

Membrane material	Type of data	Isopropanol concentration in feed (wt %)	Temperature (°C)	Permeate pressure (kPa)	Separation factor (α_{WE})	Flux (kg/m²·hr)	Comments	Ref.
Polyvinyl alcohol/ polyacrylonitrile	J_T vs. X_W	75–97	80	2	Moderate to high	~0–1.7	Highest α at low IPA concentration. PP is permeate side pressure.	Rautenbach (1988)
	Y_W vs. X_W	20–90						
	J_T vs. PP	90–95	80, 100	2–20				
	Y_W vs. PP	90–95		2–20				

Table 28 Acetone/Water Reverse Osmosis Data: Acetone-Rejecting Membranes

Membrane material	Type of data	Acetone concentration in feed (wt %)	Temperature (°C)	Feed pressure (MPa)	Acetone rejection (%)	Flux (kg/m²·hr)	Comments	Ref.
Polyacrylamide/ polysulfone	J_T, R_A	1.0	25	6.0	78–95	9.6		Dick (1982)
Polyamide (FT-30)	J_T vs. HR R_A vs. HR	0.29	30	3.45–6.89	60–90	3–36	Data from feed containing acetone, butanol, acetate, butyrate, and water. HR is hydraulic recovery.	Garcia (1986)
Polyvinyl alcohol/ polystyrene- sulfonic acid	R_A	0.02	25	7.88	40–50	NR		Koyama (1982)

Table 29 Butanol/Water Reverse Osmosis Data: Butanol-Rejecting Membranes

Membrane material	Type of data	Butanol concentration in feed (wt %)	Temperature (°C)	Feed pressure (MPa)	Butanol rejection (%)	Flux (kg/m²·hr)	Comments	Ref.
Polyacrylamide/polysulfone	J_T, R_B	1.8	25	6.0	90–96	8.3		Dick (1982)
Polyamide (FT-30)	J_T vs.HR, R_B vs. HR	0.8	30	3.45–6.89	80–89	3–36	HR is hydraulic recovery. Data from feed containing acetone, butanol, acetate, butyrate, and water.	Garcia (1986)
Polyvinyl alcohol/polystyrene-sulfonic acid	R_B	0.02	25	7.88	69	NR		Koyama (1982)
Cellulose acetate	J_T, R_B	5	25	0.5	~16	~8		Alfani (1980)
Cellulose acetate	P_E vs. FP, R_B vs. FP	0.06	25	3.0–6.1	8–15	NR	P_e is permeability. Composite data for several different membranes of the same type. FP is feed side pressure.	Kurokawa (1985)
Cellulose acetate	J_T vs. X_B, R_B vs. X_B	0.01–0.04	24 ± 1	1.72	10–20	19–32		Matsuura (1977)

Table 30 Isopropanol (IPA)/Water Reverse Osmosis Data: Isopropanol-Rejecting Membranes

Membrane material	Type of data	Ethanol concentration in feed (wt %)	Temperature (°C)	Feed pressure (MPa)	Isopropanol rejection (%)	Flux (kg/m²·hr)	Comments	Ref.
Cellulose acetate	J_T vs. X_P, R_P vs. X_P	0.006–0.03	24 ± 1	1.72	18–22	19–32		Matsuura (1977)
Cellulose acetate	J_T vs. X_P, R_P vs. X_P, J_T vs. FP, R_P vs. FP	1–15	NR	2.0–6.0			FP is feed side pressure.	Ohya (1983b)
Cellulose acetate	J_T, R_B, R_P	1, 1	25	6.0, 4.9	~6.0, 53	~65, NR		Murakami (1981)
Polyfurane/polycyanurate (PEC-1000)	J_T, R_B	1	NR	6.0	~99.6	~54		Ohya (1983b)
Polyethylenimine/m-tolylene-2,4-diisocyanate on polysulfone	J_T, R_B	1	NR	6.0	~99	~65	Essentially the NS-100 membrane.	Ohya (1983b)
Polybenzimidazolone	R_P	1	25	4.9	53	NR		Murakami (1981)

continuous fermentation without extraction (0.65 g/liter·hr). Solvent leakage was not observed over 140 hr of operation.

Product/Organic Separations

Solvent extraction and extractive fermentation are also being studied for recovery of acetone, butanol, isopropanol, and ethanol from ABE and IBE fermentations, e.g., Shimaizu and Matsubara, (1987). PV, VPe, and RO may be useful for separation of the products from the extraction solvents.

Acetone/butanol/organic PV data are presented in Table 31. Isopropanol/organic PV data are presented in Table 32. PVA membranes exhibit high separation factors for isopropanol, but exhibit low flux. Polypropylene exhibits higher flux, but lower separation factors. Relevant data for VPe and RO have not been published to date.

MISCELLANEOUS BIOFUELS AND BIOCHEMICALS

Membrane unit operations can be integrated into the recovery processes of most biofuels and biochemicals. The organic chemicals that can be produced by fermentation for which membrane separation data are available include acetaldehyde, acetic acid, butyric acid, citric acid, ethyl acetate, fusel oils, glycerol, lactic acid, propionic acid, succinic acid, and tartaric acid. Acetaldehyde production by fermentation is discussed by Armstrong et al. (1984, 1986) and Wecker and Zall (1987a, 1987b). Acetic acid production by fermentation is reviewed by Atkinson and Mavituna (1983) and Ghose and Bhadra (1985). Butyric acid production is reviewed by Playne (1985). Citric acid production is reviewed by Bouchard and Merritt (1978) and Milsom and Meers (1985). Ethyl acetate production is discussed by Armstrong et al. (1984, 1986). Glycerol production is discussed by Vijaikishore and Karanth (1984). Lactic acid production is reviewed by Atkinson and Mavituna (1983) and Vickroy (1985). Propionic acid production is reviewed by Atkinson and Mavituna (1983) and Playne (1985). Production of succinic and tartaric acids by fermentation is reviewed by Atkinson and Mavituna (1983).

Pervaporation

PV data for aqueous solutions of acetic acid, ethyl acetate, and fusel oils are presented in Table 33. Polyacrylate membranes may be useful for acetic acid dehydration. Silicone rubber may be useful for ethyl acetate removal from water. PVA exhibits high separation factors for water at low water concentrations and, hence, may be useful as an ethyl acetate dehydration membrane. The data for fusel oils demonstrate that PV can be used to dehydrate multicomponent organic streams. As additional data become available, the utility of PV in recovery of a wide range of organic chemicals produced by fermentation should become increasingly apparent.

Table 31 Acetone/Butanol/Organic Pervaporation Data: Miscellaneous Membranes

Feed mixture and membrane material	Type of data	Organic concentration in feed (wt %)	Temperature (°C)	Permeate pressure (kPa)	Separation factor (α_{AB})	Flux (kg/m²·hr)	Comments	Ref.
Acetone/Butanol								
Polypropylene	J_i vs. PP J_i vs. T	$X_A = 0.5$ $X_B = 1.0$	30 15–50	1–31 2.0	Comments	0.001–0.025	Acetone is component A and butanol is component B; butanol is the more permeable component (i.e., $J_B \sim 10 J_A$).	Ohya (1988)
Butanol/Cyclo-hexane								
PTFE/PVP[a]	J_T, α	$X_A = 10$ $X_B = 90$	25	~0	23.5	0.3	Butanol is component A, i.e., the more permeable component.	Aptel (1976)

[a]PTFE/PVP is polytetrafluoroethylene/poly-N-vinylpyrrolidone.

Table 32 Isopropanol (IPA)/Organic Pervaporation Data

Feed mixture and membrane material	Type of data	Isopropanol concentration in feed (wt %)	Temperature (°C)	Permeate pressure (kPa)	Separation factor (α_{AB})	Flux (kg/m²·hr)	Comments	Ref.
IPA/n-Hexane Polyvinyl alcohol (GFT)	J_T, α	10	NR	NR	~1500	0.016	Isopropanol (IPA) is component A, i.e., the component that preferentially permeates.	Fleming (1989a)
Polyvinyl alcohol	J_T, Y_P	10	60	NR	>900	<0.01	Isopropanol is component A.	Schneider (1987)
IPA/Toluene Polyvinyl alcohol (GFT)	J_T, α	10	NR	NR	~2200	~0.026	Isopropanol is component A.	Fleming (1989a)
Polyvinyl alcohol	J_T, Y_P	10	60	NR	>900	<0.02	Isopropanol is component A. α calculated from Y_P	Schneider (1987)
IPA/Benzene Polyethylene	J_T vs. X_B Y_B vs. X_B	29–70	42–60	NR	Moderate to low	0.1–2.0	Benzene is component A., i.e., the component that preferentially permeates.	Carter (1964)

Table 33 Miscellaneous Organics Pervaporation Data

Feed mixture and membrane material	Type of data	Organic concentration in feed (wt %)	Temperature (°C)	Permeate pressure (kPa)
Acetic Acid/Water				
PTMSP	J_T vs. X_A	10–85	30	0.01
	α vs. X_A			
	Y_A vs. X_A			
Polyether-block-amides	J_T vs. X_A	1.5–9.0	50	<0.2
	Y_A vs. X_A	0–10		
Silicone rubber	J_T, α	1.1	50	<0.2
Polyacrylic acid	J_T, α	48	15	NAR
Polyacrylic acid/nylon-6 blend	PR vs. X_B	24–97	15–35	NAR
	PR vs. T			
	α vs. X_B			
	α vs. T			
Ethyl Acetate/Water				
Silicone rubber (PDMS)	Y_A vs. X_A	0–4.0	30	2–4
Polyvinyl alcohol (GFT)	Y_B vs. X_B	0–100	60	3
	Y_B vs. X_B	0–1.0	60	3
Polyolefin/PDMS	Y_A vs. X_A	0–10	30	2–4
Polyolefin	Y_A vs. X_A	0–1.0	30	2–4
Fusel Oils/Water				
GFT dehydration membrane	J_A vs. X_A	87–100		
	α vs. X_A	85–100	90	NR
	Y_B vs. X_B	85–100		
GFT membrane for ethanol permeation	J_A vs. X_A	87–100		
	α vs. X_B	85–100	60, 90	NR
	Y_B vs. X_B	85–100		

Separation factor (α_{AB})	Flux (kg/m^2·hr)	Comments	Ref.
3–6	0.25–0.35	Acetic acid is component A; water is B. PTMSP is poly[1-(trimethylsilyl)-1-propyne]. Membrane thickness is ~30 μm. Maximum α for 25 < X_A < 50 wt % acetic acid.	Masuda (1990)
Low	0.18–0.28	Acetic acid is component A. Water is component B.	Bengston (1988)
1.25	0.39	Water is component A. Acetic acid is component B.	Bengston (1988)
2–8	0.4–0.55	Water is component A; acetic acid is B. NAR: Not adequately reported.	Huang (1987)
22–82	NAR	Water is component A; acetic acid is B. Highest α at high acetic acid concentrations. NAR: Not adequately reported.	Huang (1987)
Moderate to high	NR	Membrane thickness is 1.0 μm. Ethyl acetate is component A; water is B. PDMS is polydimethylsiloxane (silicone rubber).	Blume (1987) Blume (1990)
Low to high	NR NR	Water is component A; ethyl acetate is B. α is highest at low water concentrations.	Wesslein (1988, 1990)
High	NR	Membrane thickness: 3.5 μm. Ethyl acetate is component A; water is B.	Blume (1990)
Very high	NR	Ethyl acetate is component A; water is B.	Blume (1990)
50–500	~0–2.3	Water is component A; fusel oils are B. Composition of synthetic fusel oils: Water 13.1% 3-methyl-1-butanol 64.1% Isobutanol 12.5% n-Butanol 0.65% n-Propanol 0.84% Ethanol 8.81%	Kraetz (1988)
150–10,000	~0–1.6	Water is component A; fusel oils are B. Composition of synthetic fusel oils given above.	Kraetz (1988)

Reverse Osmosis

RO data for aqueous solutions of acetic acid, butyric acid, citric acid, glycerol, propionic acid, succinic acid, and tartaric acid are presented in Table 34. High acetate and butyrate rejections are obtained with the polyamide membrane, which also exhibits high flux. The polybenzimidazolone (PBIL) membrane exhibits high rejections for citric acid, glycerol, succinic acid, and tartaric acid. Cellulose acetate exhibits low glycerol rejection. Desai (1972) studied RO separation of glycerol from polyglycerins and salts; overall glycerol rejection with cellulose acetate was about 22%.

Lactic acid recovery from fermentation broth using RO has been studied by Smith et al. (1977); fouling of the membranes was a significant problem. More recently, RO was studied for this separation by Schlicher and Cheryan (1990). Two membranes were studied: cellulose acetate and a thin-film composite (PCI ZF-99). The thin-film composite exhibited the best performance. Using fermentation broths (pH 6.0) containing 0.5–4.4 wt % lactic acid as the feed, the flux obtained with this membrane varied between 0 and 45 KMH and lactic acid rejection was 90–99.5%. The authors conclude that, with the use of high flow velocities, concentration of lactic acid from 1.0 to 12.0 wt % should be possible at feed pressures of 6.9 MPa. The energy required by this operation is $\sim$54 MJ/m^3.

Electrodialysis (ED)

The recovery of lactic acid from fermentation broth by use of ED bioreactors has been investigated as a means to continuously recover the acid, to reduce feedback inhibition, and to control pH without the addition of base. The process concept is shown in Figure 9.

In an early study by Kumphanzl and Dye (1962), fouling of the anion exchange membrane by cells and other macromolecules in the broth adversely affected performance. In a more recent study, Nomura et al. (1987) also encountered fouling problems due to cell adsorption to the membranes. They coupled ED to an immobilized cell bioreactor and operated the system for 8 months at improved productivity as compared to batch fermentation and a cell-free ED bioreactor. In other recent studies (Hongo et al., 1986, and Boyaval et al., 1987), productivity increases of three- to fourfold are reported. Rosenau et al. (1986) obtained 90% recovery of lactic acid from broth using ED. In studies by de Raucourt et al. (1989a, 1989b), broth was first ultrafiltered prior to treatment by ED with encouraging results; the papers by de Raucourt et al. consist primarily of process modeling.

Acetic acid recovery by continuous ED/fermentation has been investigated by Nomura et al. (1988, 1989). In this study, the substrate for acetic acid synthesis was ethanol. The system was operated for 30 days. A volumetric productivity of $\sim$5.5 g acetic acid/liter·hr was reported, an improvement of about 35–70% over conventional continuous fermentation.

Membrane Extraction

The extraction of ethyl acetate, acetic acid, and ethanol from water using gas membranes was studied by Qin et al. (1990). In this concept, a gas is suspended in hydrophobic, microporous hollow fibers. Organic compounds in an aqueous feedstream volatilize into the gas and diffuse across the membrane. The permeate side contains a stream containing sodium hydroxide, into which the organic compounds dissolve. Candidate gases include sulfur dioxide, bromine, hydrogen sulfide, and hydrogen cyanide. The mass transfer coefficients observed for ethyl acetate and ethanol were much larger than that observed for acetic acid.

CONCLUSIONS

A tremendous number of data have been generated since 1985 on the use of membranes for the recovery of organic chemicals. Many of these chemicals can be produced by fermentation. However, the literature on the recovery of organic chemicals from fermentation broth could use some focusing.

First, for most membranes, additional detailed performance data are needed in the ranges of potential application. For instance, consider the separation of ethanol from water. Many typical studies on membrane separation of ethanol and water report performance as a function of feed ethanol concentration at 10, 30, 50, 70, and 90 wt % ethanol. Yet the range of 10–90 wt % is the concentration range in which membrane operations have the least likelihood of being used. Conventional fractional distillation is extremely efficient in this range and is the most likely operation to be used for concentrating ethanol between 10 and 90 wt %.

Membranes are most likely to be used to preconcentrate ethanol in the range from 0 to 10 wt %, followed by fractional distillation to produce ethanol near its azeotropic concentration (95.56 wt %), and for dehydrating ethanol (85–99.5 wt % range) after fractional distillation has been used to produce concentrated ethanol. Consider Figure 10. The energy requirements of fractional distillation are highly dependent on feed concentration for feeds containing less than 10 wt % ethanol. Between 10 and 20 wt % ethanol, the energy requirements become nearly independent of feed concentration. Membrane operations have low energy-related operating costs due to low energy requirements. The concentration ranges in which distillation has high energy requirements are the ranges in which membranes are most likely to have an advantage over distillation. Therefore, membrane operations are most likely to be used when the feed is extremely dilute (e.g., 1–6 wt % ethanol). Membrane systems are not likely to have an advantage for producing 20–80 wt % ethanol.

On the other end of the concentration range, the conventional process using fractional distillation, followed by azeotropic distillation, for dehydrating ethanol (Figure 11) is energy-intensive. This process has high energy requirements because the optimum ethanol feed concentration for azeotropic distillation is close

Table 34 Reverse Osmosis Data for Miscellaneous Organic/Water Mixtures

Feed mixture and membrane material	Type of data	Organic concentration in feed (wt %)	Temperature (°C)	Feed pressure (MPa)
Acetaldehyde/Water				
Polybenzimidalozone	R_A	1.0	25	4.9
Cellulose acetate	R_A	1.0	25	4.9
Acetic Acid/Water				
Polyvinyl alcohol/poly-styrenesulfonic acid	R_{AA}	0.02	25	7.88
Polyamide (FT-30)	J_T vs. HR R_{AA} vs. HR	0.07	30	3.45–6.89
Cellulose acetate	R_{AA}	1.0	25	4.9
Polybenzimidazolone	R_{AA}	1.0	25	4.9
Butyric Acid/Water				
Polyamide (FT-30)	R_B vs. HR J_T vs. HR	0.09	30	3.45–6.89
Citric Acid/Water				
Polybenzimidazolone	R_{CA}	1.0	25	4.9
Glycerol/Water				
Cellulose acetate	J_T vs. X_G	2.5–10	25	4.14
Cellulose acetate	J_T vs. X_G R_G vs. X_G	2.5–10	25	4.14
Polybenzimidazolone	R_G	1.0	25	4.9
Propionic Acid/Water				
Polyvinyl alcohol/poly-styrenesulfonic acid	R_{PA}	0.02	25	7.88
Polybenzimidazolone	R_{PA}	0.02	25	4.9
Succinic Acid/Water				
Polybenzimidazolone	R_{SA}	1.0	25	4.9
Tartaric Acid/Water				
Polybenzimidazolone	R_{TA}	1.0	25	4.9

Organic rejection (%)	Flux (kg/m²·hr)	Comments	Ref.
50	NR	Acetaldehyde (A) is rejected.	Murakami (1981)
23	NR	Acetaldehyde is rejected.	Murakami (1981)
~60	NR	Acetic acid (AA) is rejected.	Koyama (1982)
~92	3–36	Acetic acid is rejected. Feed to this test contained acetone, butanol, acetate, butyrate, and water. HR is hydraulic retention.	Garcia (1986)
7	NR	Acetic acid is rejected.	Murakami (1981)
40	NR	Acetic acid is rejected.	Murakami (1981)
~95	3–36	Butyric (B) acid is rejected. Feed to this test contained acetone, butanol, acetate, butyrate, and water. HR is hydraulic retention.	Garcia (1986)
99	NR	Citric acid (CA) is rejected.	Murakami (1981)
20–40	2.9–8.1	Glycerol (G) is rejected. Feed was aqueous solution of glycerol.	Vijaikishore (1981)
10–20	1.6–6.5	Glycerol is rejected. Feed was clarified, diluted fermentation broth.	Vijaikishore (1987)
97	NR	Glycerol is rejected.	Murakami (1981)
~70	NR	Propionic acid (PA) is rejected.	Koyama (1982)
67	NR	Propionic acid is rejected.	Murakami (1981)
95	NR	Succinic acid (SA) is rejected.	Murakami (1981)
99	NR	Tartaric acid (TA) is rejected.	Murakami (1981)

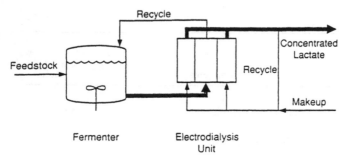

Figure 9 The electrodialysis bioreactor concept.

to the azeotropic concentration of ethanol/water solutions (see Figure 12). The energy requirements of fractional distillation increase dramatically as the distillate approaches the azeotropic concentration. Membrane systems, on the other hand, can be used to break the azeotrope without requiring feedstreams near the azeotropic concentration. For this reason, membrane systems have a good opportunity to compete against conventional distillation for dehydrating ethanol. Pervaporation and vapor permeation are already emerging as viable alternatives to the conventional ethanol dehydration processes.

The point of this discussion is the following: In order to make membrane

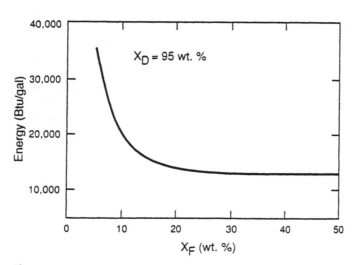

Figure 10 Ethanol recovery: energy requirements of fractional distillation as a function of feed concentration.

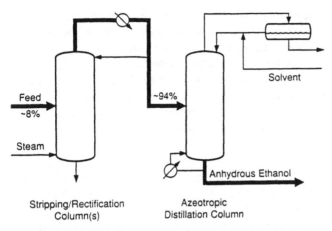

Figure 11 Conventional distillation process for dehydration of ethanol.

performance data more useful to the process design engineer, studies should focus on the concentration ranges in which the membrane has potential to be used. Very little can actually be learned about the potential of a new membrane material for the recovery of ethanol, for instance, from data in the 10–90 wt % range. Detailed data in the 0–10 wt % and 85–100 wt % range are needed to assess the potential of

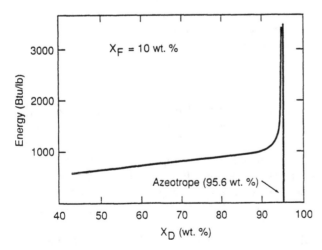

Figure 12 Ethanol recovery: energy requirements of fractional distillation as a function of distillate concentration.

a membrane for recovery of ethanol. After studying membrane performance data on a specific chemical and analyzing the conventional processes for recovery of a specific chemical, authors can determine the concentration ranges in which membranes may have potential; it would be helpful if detailed data were reported in the concentration ranges of potential application. Data should be reported as performance vs. retentate side concentration (for several data points), feed or permeate side pressure, flow velocity, etc. McCabe–Thiele–type diagrams of permeate side concentration vs. feed side concentration for membranes and modules are increasingly being used for reporting membrane performance data; these diagrams are extremely useful and this trend toward their use should continue. In addition, the distinction as to whether the data are determined using the integral and differential method should be clearly made (Leeper, 1986).

Regarding focusing of the literature, the second point is that additional process analysis studies are needed. A common basis for comparison of alternate processes can emerge from such analyses. These analyses can then be used to provide the information needed to determine the best process alternatives for recovery of specific chemicals.

In conclusion, membrane operations are emerging as a viable candidates for integration into processes for production and recovery of biofuels and biochemicals. Pervaporation and vapor permeation are presently nearing commercial status. Pervaporation, vapor permeation, reverse osmosis, electrodialysis, and membrane extraction should be considered when developing processes for recovery of organic chemicals. The present commercial applications of membrane operations in the recovery of chemicals are hybrid processes, which take advantage of the specific separation capabilities of membranes. When considering the use of membrane separations, the possibility of combining separation operations into hybrid processes should be strongly considered so that membranes can be used to full advantage.

ACKNOWLEDGMENTS

This chapter was supported under contract no. DE-AC07-76ID01570 from the U.S. Department of Energy (DOE), to the Idaho National Engineering Laboratory/ EG&G Idaho, Inc., for DOE Office of Industrial Technology (DOE-OIT). The author is grateful to the following persons for their help and insightful comments during the preparation of this chapter: Dr. Richard Baker (MTR, Inc., Menlo Park, CA), Dr. Hugh Fleming (Zenon Environmental Group, Inc., Sussex, NJ), and Dr. Teh-An Hsu (EG&G Idaho, Inc., Idaho Falls, ID, currently, National Renewable Energy Laboratory, Golden, CO). The support of Mr. William M. Sonnett (DOE-OIT, Washington, D.C.), Mr. David M. Blanchfield (DOE–Idaho Operations Office, Idaho Falls, ID), and Mr. Thomas Lawford (EG&G Idaho, Inc., Idaho Falls, ID) is also deeply appreciated.

REFERENCES

Anonymous (1989), Facts and Figures for the Chemical Industry, *Chem. Eng. News 67*(25), 36 (June 19).

Alfani, F., et al. (1980), Effect of alcohols on the mechanical and transport properties of asymmetric acetate membranes. *Water Res. 14*, 461.

Aptel, P., et al. (1976), Application of the pervaporation process to separate azeotropic mixtures. *J. Membrane Sci. 1*, 271.

Armstrong, D. W., et al. (1984), Production of ethyl acetate from dilute ethanol solutions by *Candida utilis. Biotechnol. Bioeng. 26*(9), 1038–1041.

Armstrong, D. W., et al. (1986), Biological production of economically recoverable products from dilute ethanol streams. In *Biotechnology and Renewable Energy* (M. Moo-Young et al., eds.), Elsevier, New York, pp. 153–160.

Asada, T. (1988), Dehydration of organic solvents: Some actual results of pervaporation plants in Japan. In *Proceedings of Third International Conference on Pervaporation Processes in the Chemical Industry, Nancy, France, September 19–22, 1988* (R. Bakish, ed.), Bakish Materials Corporation, Englewood, NJ, pp. 379–386.

Atkinson, B., and F. Mavituna (1983), *Biochemical Engineering and Biotechnology Handbook*, Nature Press, New York.

Awang, G. M., et al. (1988), The acetone-butanol-ethanol fermentation. *CRC Crit. Rev. Biotechnol. 15* (Supplement 1), S33–S67.

Bandini, S., et al. (1988), Vacuum membrane distillation: Pervaporation through porous hydrophobic membranes. In *Proceedings of Third International Conference on Pervaporation Processes in the Chemical Industry, Nancy, France, September 19–22, 1988* (R. Bakish, ed.), Bakish Materials Corporation, Englewood, NJ, pp. 117–126.

Bartels, C. R., et al. (1988), Plant evaluation of pervaporation process. In *Proceedings of Third International Conference on Pervaporation Processes in the Chemical Industry, Nancy, France, September 19–22, 1988* (R. Bakish, ed.), Bakish Materials Corporation, Englewood, NJ, pp. 486–492.

Beaver, Earl (1989), Permea, Inc. (Monsanto), St. Louis, MO, personal communication, October 19.

Beesch, S. C. (1952), Acetone-butanol fermentation of sugars. *Indust. Eng. Chem. 44*(7), 1677–1682.

Belanche, M. I., and J. Lora (1988), Pervaporation selectivity changes with material of membranes. In *Proceedings of Third International Conference on Pervaporation Processes in the Chemical Industry, Nancy, France, September 19–22, 1988* (R. Bakish, ed.), Bakish Materials Corporation, Englewood, NJ, pp. 150–157.

Bengston, G., and K. W. Boddeker (1988), Pervaporation of low volatiles from water. In *Proceedings of Third International Conference on Pervaporation Processes in the Chemical Industry, Nancy, France, September 19–22, 1988* (R. Bakish, ed.), Bakish Materials Corporation, Englewood, NJ, pp. 439–448.

Bitter, J. G. A. (1988), Evaluation of membrane technology for separation of dilute alcohol solutions. In *Proceedings of Third International Conference on Pervaporation Processes in the Chemical Industry, Nancy, France, September 19–22, 1988* (R. Bakish, ed.), Bakish Materials Corporation, Englewood, NJ, pp. 476–485.

Blume, I., and R. Baker (1987), Separation and concentration of organic solvents from water using pervaporation. In *Proceedings of Second International Conference on Pervaporation Processes in the Chemical Industry, San Antonio, TX, March 8–11, 1987* (R. Bakish, ed.), Bakish Materials Corporation, Englewood, NJ, pp. 111–125.

Blume, I., et al. (1990). The separation of dissolved organics from water by pervaporation. *J. Membrane Sci. 49*(3), 253–286 (April 15).

K. W. Boddeker (1990), Terminology in pervaporation. *J. Membrane Sci. 51*(3), 259–272 (August 1).

Bouchard, E. F., and E. G. Merritt (1978), Citric acid. In *Kirk-Othmer Encyclopedia of Chemical Technology, 3rd Ed.*, Vol. 6, John Wiley and Sons, New York, pp. 150–179.

Boyaval, P., et al. (1987), Continuous lactic acid fermentation with concentrated product recovery by ultrafiltration and electrodialysis. *Biotechnol. Lett. 9*(3), 207–212.

Bruschke, H. E. (1988), State of art of pervaporation. In *Proceedings of Third International Conference on Pervaporation Processes in the Chemical Industry, Nancy, France, September 19–22, 1988* (R. Bakish, ed.), Bakish Materials Corporation, Englewood, NJ, pp. 2–11.

Bruschke, H. E. (1990a), State of the art. In *Proceedings of Fourth International Conference on Pervaporation Processes in the Chemical Industry, Ft. Lauderdale, FL, December 3–7, 1989* (R. Bakish, ed.), Bakish Materials Corporation, Englewood, NJ, p. 2.

Bruschke, H. E. A. (1990b), Removal of ethanol from aqueous streams by pervaporation. *Desalination 77*, 323–329.

Busche, R. M. (1989), The biomass alternative: A national insurance policy to protect the US strategic supply of chemical feedstocks. *Appl. Biochem. Biotechnol. 20/21*, 665–674.

Cabasso, I., et al. (1974), Permeation of organic solvents through polymeric membranes based on polymeric alloys of polyphosphonates and acetyl cellulose. II. Separation of benzene, cyclohexene, and cyclohexane. *J. Appl. Poly. Sci. 18*(7), 2137–2147.

Cabasso, I., and Z. Z. Liu (1985), The permselectivity of ion-exchange membranes for non-electrolyte liquid mixtures. I. Separation of alcohol/water mixtures with Nafion hollow fibers. *J. Membrane Sci. 24*(1),101–119.

Cabasso, I., et al. (1985), On the separation of alcohol/water mixtures by polyethylene ion exchange membranes. *J. Polym. Sci.: Poly. Lett. Ed. 23*, 577–581.

Cabasso, I., et al. (1986), The permselectivity of ion-exchange membranes for non-electrolyte liquid mixtures. II. The effect of counterions (separations of alcohol/water mixtures with Nafion membranes). *J. Membrane Sci. 28*, 109–122.

Calibo, R. L., et al. (1987), Ethanol stripping by pervaporation using porous PTFE membrane. *J. Ferm. Technol. 65*(6), 665–674.

Calibo, R. L., et al. (1989), Continuous ethanol fermentation of concentrated sugar solutions coupled with membrane distillation using a PTFE module. *J. Ferm. Bioeng. 67*(1), 40–45.

Carter, J. W., and B. Jagannadhaswamy (1964), Separation of organic liquids by selective permeation through polymeric films. *Br. Chem. Eng. 9*(8), 523–526.

Changlou, Z., et al. (1987), Separation of ethanol-water mixtures by pervaporation— Membrane separation process. *Desalination 62*, 299–313.

Changlou, Z., et al. (1988), Description of Pvap mechanism with separation-characteristics graphs. In *Proceedings of Third International Conference on Pervaporation Processes in the Chemical Industry, Nancy, France, September 19–22, 1988* (R. Bakish, ed.), Bakish Materials Corporation, Englewood, NJ, pp. 44–53.

Changlou, Z., et al. (1989), A study on characteristics and enhancement of pervaporation—Membrane separation process. *Desalination 71*(1), 1–18.

Cheryan, M. (1986), *Ultrafiltration Handbook*, Technomic, Lancaster, PA.

Choudhury, J. P., et al. (1985), Separation of ethanol from ethanol-water mixtures by reverse osmosis, *Biotechnol. Bioeng. 27*(7), 1081–1084.

Choudhury, J. P., and P. Ghosh (1986), Ethanol separation from molasses based fermentation broth by reverse osmosis. *Biotechnol. Lett. 8*(10), 731–734.

Choudhury, J. P., et al. (1988), Styrene-grafted cellulose acetate reverse osmosis membrane for ethanol separation. *J. Membrane Sci. 35*(3), 301–310.

Cogat, P. O. (1988), Dehydration of ethanol: Pervaporation compared to azeotropic distillation. In *Proceedings of Third International Conference on Pervaporation Processes in the Chemical Industry, Nancy, France, September 19–22, 1988* (R. Bakish, ed.), Bakish Materials Corporation, Englewood, NJ, pp. 305–316.

Collura, M. A., and W. L. Luyben (1988), Energy-saving distillation designs in ethanol production. *Indust. Eng. Chem. Res. 27*, 1686–1696.

Crabbe, P. G., et al. (1986), Effect of microorganisms on rate of liquid extraction of ethanol from fermentation broths. *Biotechnol. Bioeng. 28*, 939–943.

de Pinho, M., et al. (1988), Reverse osmosis separation of glucose-ethanol-water system by cellulose acetate membranes. *Chem. Eng. Commun. 64*, 113–123.

de Raucourt, A., et al. (1989a), Lactose continuous fermentation with cells recycled by ultrafiltration and lactate separation by electrodialysis: Modelling and simulation. *Appl. Microbiol. Biotechnol. 30*, 521–527.

de Raucourt, A., et al. (1989b), Lactose continuous fermentation with cells recycled by ultrafiltration and lactate separation by electrodialysis: Model identification. *Appl. Microbiol. Biotechnol. 30*, 528–534.

Desai, S. V., et al. (1972), An economically attractive application of reverse osmosis to refinement of a petrochemical effluent stream, *Water—1972 (AIChE Symposium Series Number 124, Volume 68)* (L. K. Cecil, ed.), American Institute of Chemical Engineers, New York, pp. 379–387.

Dick, R., and G. Mavel (1982), R. O. technology for preconcentrating alcoholic solutions. In *Proceedings, Fifth International Alcohol Fuel Technology Symposium, Volume I, Auckland, New Zealand, May 13–18, 1982*, John McIndoe Ltd., Dunedin, New Zealand, pp. 151–158.

DOE (1990), *Membrane Separation Systems: A Research Needs Assessment. Volume 1. Executive Summary. Volume 2. Final Report*, DOE Report No. DOE/ER/30133-H1.

Egan, B. Z., et al. (1988), Solvent extraction and recovery of ethanol from aqueous solutions. *Indust. Eng. Chem. Res. 27*(7), 1330–1332.

Ellinghorst, G., et al. (1987), Membranes of pervaporation by radiation grafting and curing and by plasma processing. In *Proceedings of Second International Conference on Pervaporation Processes in the Chemical Industry, San Antonio, TX, March 8–11, 1987* (R. Bakish, ed.), Bakish Materials Corporation, Englewood, NJ, pp. 79–99.

Ennis, B. M., et al. (1986), The acetone-butanol-ethanol fermentation: A current assessment. *Process Biochem. 21*(5), 131–147.

Eykamp, W., and J. Steen (1987), Ultrafiltration and reverse osmosis. In *Handbook of Separation Technology* (R. W. Rosseau, ed.), John Wiley and Sons, New York, pp. 826–839.

Featherstone, W., and T. Cox (1971), Separation of aqueous-organic mixtures by pervaporation. *BCE Process Technol. 16*, 817.

Fleming, H. L. (1989a), Membrane pervaporation: Separation of organic aqueous mixtures, *International Conference on Fuel Alcohols and Chemicals*, Guadalajara, Mexico, February 2, 1989.

Fleming, H. L. (1989b), Membrane pervaporation: Separation of organic aqueous mixtures. *Sixth Symposium on Separation Science and Technology for Energy Applications*, Knoxville, TN, October 22–26, 1989.

Fournier, R. L. (1986), Mathematical model of extractive fermentation: Application to the production of ethanol. *Biotechnol. Bioeng. 28*(8), 1206–1212.

Frennesson, I., et al. (1986), Pervaporation and ethanol upgrading—A literature review. *Chem. Eng. Commun. 45*, 277–289.

Garcia, A., III, et al. (1986), Butanol fermentation liquor production and separation by reverse osmosis. *Biotechnol. Bioeng. 28*(6), 785–791.

Ghose, T. K., and A. Bhadra (1985), Acetic acid. *Comprehensive Biotechnology, Vol. 3, The Practice of Biotechnology: Current Commodity Products* (M. Moo-Young et al., eds.), Pergamon Press, New York, pp. 701–729.

Gooding, C. H., and F. J. Bahouth (1985), Membrane-aided distillation of azeotropic solutions. *Chem. Eng. Commun. 35*, 267–279.

Gooding, C. H. (1986), *Pervaporative Dehydration of Ethanol, Annual Report*, DOE Report No. DOE/CE/30792-T22, NTIS Order No. DE86008070.

Gregor, H. P., and J. W. Jeffries (1979), *Ethanolic Fuels from Renewable Resources in the Solar Age*, NTIS Report No. PB-29564.

Groot, W. J., et al. (1984a), Pervaporation for simultaneous product recovery in the butanol/isopropanol batch fermentation. *Biotechnol. Lett. 6*(11), 709–714.

Groot, W. J., et al. (1984b), Increase of substrate conversion by pervaporation in the continuous butanol fermentation. *Biotechnol. Lett. 6*(12), 789–792.

Groot, W. J., and Ch. A. M. Luyben (1987), Continuous production of butanol from a glucose/xylose mixture with and immobilized cell system coupled to pervaporation. *Biotechnol. Lett. 9*(12), 867–870.

Groot, W. J., et al. (1988), Pervaporation of fermentation products: Mass transfer of solutes in silicone membranes and the performance of pervaporation in a fermentation. In *Proceedings of Third International Conference on Pervaporation Processes in the Chemical Industry, Nancy, France, September 19–22, 1988* (R. Bakish, ed.), Bakish Materials Corporation, Englewood, NJ, pp. 398–404.

Guanwen, C., and L. Fengcai (1988), Pervaporation of organic liquid mixtures through polyphenylquinoxaline membranes. In *Proceedings of Third International Conference on Pervaporation Processes in the Chemical Industry, Nancy, France, September 19–22, 1988* (R. Bakish, ed.), Bakish Materials Corporation, Englewood, NJ, pp. 188–193.

Haggstrom, L. (1985), Acetone-butanol fermentation and its variants. *Biotechnol. Adv. 3*(1), 13–28.

Hagwood, B. (1989), Information Resources, Inc. Washington, D.C., personal communication.

Hallstrom, B., and G. von Sengbusch (1986), *Terminology for Pressure Driven Membrane Operations*, issued by the European Society of Membrane Science and Technology. Available at no charge from Dr. Enrico Drioli, University of Naples, Department of Chemical Engineering, Piazzale Tecchio, Naples, Italy I 80125.

Hauser, J., et al. (1987), Sorption, diffusion and pervaporation of water/alcohol-mixtures in PVA membranes. Experimental results and theoretical treatment. In *Proceedings of Second International Conference on Pervaporation Processes in the Chemical Industry, San Antonio, TX, March 8–11, 1987* (R. Bakish, ed.), Bakish Materials Corporation, Englewood, NJ, pp. 15–34.

Hauser, J., et al. (1988), Non-ideal behavior of solubilities of liquid mixtures in PVA and its influence on pervaporation. *Proceedings of Third International Conference on Pervaporation Processes in the Chemical Industry, Nancy, France, September 19–22, 1988* (R. Bakish, ed.), Bakish Materials Corporation, Englewood, NJ, pp. 15–34.

Hauser, J., et al. (1989), Non-ideal solubility of liquid mixtures in poly(vinyl alcohol) and its influence on pervaporation. *J. Membrane Sci. 47*, 261–276.

Hirotsu, T. (1987a), Water-ethanol separation by pervaporation through plasma graft polymerized membranes. *J. Appl. Poly. Sci. 34*, 1159–1172.

Hirotsu, T. (1987b), Graft polymerized membranes of methacrylic acid by plasma for water-ethanol permseparation. *Indust. Eng. Chem. Res. 26*(7), 1287–1290.

Hirotsu, T. (1988a), Water-ethanol separation by pervaporation through plasma graft polymerized membranes of HEMA's. In *Proceedings of Third International Conference on Pervaporation Processes in the Chemical Industry, Nancy, France, September 19–22, 1988* (R. Bakish, ed.), Bakish Materials Corporation, Englewood, NJ, pp. 103–109.

Hirotsu, T. (1988b), Water-ethanol separation by pervaporation through plasma graft polymerized membranes of acrylic acid and acrylamide. In *Proceedings of Third International Conference on Pervaporation Processes in the Chemical Industry, Nancy, France, September 19–22, 1988* (R. Bakish, ed.), Bakish Materials Corporation, Englewood, NJ, pp. 110–116.

Hirotsu, T. (1989), Water-ethanol separation by pervaporation through plasma-graft-polymerized membranes of 2-hydroxyethyl methacrylate with acrylic acid and methacrylic acid. *J. Membrane Sci. 45*(1+2), 137–154.

Hoffman, E., et al. (1987), Evaporation of alcohol/water mixtures through hydrophobic porous membranes. *J. Membrane Sci. 34*(2), 199–206.

Honda, H., et al. (1987), A general framework for the assessment of extractive fermentations. *Chem. Eng. Sci. 42*(3), 493–498.

Hongo, M., et al. (1986), Novel method of lactic acid production by electrodialysis fermentation. *Appl. Env. Microbiol. 52*(2), 314–319.

Hoover, K. C., and S.-T. Hwang (1982), Pervaporation by a continuous membrane column. *J. Membrane Sci. 10*, 253 (1982).

Horsley, L. H. (ed.) (1973), *Azeotropic Data—III (Advances in Chemistry Series 116)*, American Chemical Society, Washington, D. C., p. 18.

Huang, S. Y., and W. C. Ko (1984), Concentration of ethyl alcohol aqueous solutions. *Maku (Membrane) 9*(2), 113–120.

Huang, R. Y. M., et al. (1987), Novel blended nylon membranes for the pervaporation separation of acetic acid-water and ethanol-water liquid mixture water systems. In *Proceedings of Second International Conference on Pervaporation Processes in the Chemical Industry, San Antonio, TX, March 8–11, 1987* (R. Bakish, ed.), Bakish Materials Corporation, Englewood, NJ, pp. 225–239.

Huang, R. Y. M., and C. K. Yeom (1990), Pervaporation separation of aqueous mixtures using crosslinked poly(vinyl alcohol) (PVA). II. Permeation of ethanol-water mixtures. *J. Membrane Sci. 51*(3), 273–292 (August 1).

Hwang, S.-T., and J. M. Thorman (1980), The continuous membrane column. *AIChE J. 26*, 558.

Hwang, S.-T., and K. Kammermeyer (1984), *Membranes in Separations* (Techniques of Chemistry, Volume 7), Krieger, Malabar, FL.

Inagagaki, N., et al. (1988), Pervaporation of ethanol-water mixture by plasma films prepared from hexamethyldisiloxane. *Desalination 70*, 465–479.

Ishida, M., and N. Nakagawa (1985), Exergy analysis of a pervaporation system and its combination with a distillation column based on an energy utilization diagram. *J. Membrane Sci. 24*, 274–283.

Ishihara, K., et al. (1986), Pervaporation of alcohol/water mixtures through poly[1-(tri-methylsilyl)-1-propyne] membrane. *Mackromol. Chem. Rapid Commun. 7*, 43–46.

Jansen, A. E., et al. (1988), Dehydration of alcohols by vapor permeation. In *Proceedings of Third International Conference on Pervaporation Processes in the Chemical Industry, Nancy, France, September 19–22, 1988* (R. Bakish, ed.), Bakish Materials Corporation, Englewood, NJ, pp. 338–341.

Jeffries, T. W. (1985), Emerging technology for fermenting D-xylose. *Trends Biotechnol. 3*(8), 208–212.

Jeon, Y. J., and Y. Y. Lee (1989), In situ product separation in butanol fermentation by membrane-assisted extraction. *Enzyme Microb. Technol. 11*, 575–582 (September).

Jones, D. T., and D. R. Woods (1986), Acetone-butanol fermentation revisited. *Microbiol. Rev. 50*(4), 484–524.

Kang, Y. S., et al. (1990a), Pervaporation of water-ethanol mixtures through crosslinked and surface-modified poly(vinyl alcohol) membrane. *J. Membrane Sci. 51*, 215–226.

Kang, W., et al. (1990b), Ethanol production in a microporous hollow-fiber-based extractive fermentor with immobilized yeast. *Biotechnol. Bioeng. 36*, 826–833.

Karakane, H., et al. (1988), Separation of water-ethanol by pervaporation through poly-electrolyte complex composite membranes. In *Proceedings of Third International Conference on Pervaporation Processes in the Chemical Industry, Nancy, France, September 19–22, 1988* (R. Bakish, ed.), Bakish Materials Corporation, Englewood, NJ, pp. 194–202.

Kaschemekat, J., et al. (1988), Separation of organics from water using pervaporation. In *Proceedings of Third International Conference on Pervaporation Processes in the Chemical Industry, Nancy, France, September 19–22, 1988* (R. Bakish, ed.), Bakish Materials Corporation, Englewood, NJ, pp. 405–412.

Kashiwagi, T., et al. (1988), Separation of ethanol from ethanol/water mixtures by plasma-polymerized membranes from silicone compounds. *J. Membrane Sci. 36*, 353–362.

Kimura, S., and T. Nomura (1983), Pervaporation of organic substance water system with silicone rubber membrane. *Maku (Membrane)*, 8(3), 177–183 (in Japanese with English abstract).

Klein, E., et al. (1987), Membrane processes: Dialysis and electrodialysis. In *Handbook of Separation Process Technology* (R. W. Rousseau, ed.), John Wiley and Sons, New York, pp. 954–981.

Klinkowski, P. R. (1983), Ultrafiltration. In *Kirk-Othmer Encyclopedia of Chemical Technology*, 3rd Ed., Vol. 23, John Wiley and Sons, New York, pp. 439–461.

Kollerup, F., and A. J. Daugulis (1985a), A mathematical model for ethanol production by extractive fermentation in a continuous stirred tank fermentor. *Biotechnol. Bioeng.* 27(9), 1335–1346.

Kollerup, F., and A. J. Daugulis (1985b), Screening and identification of extractive fermentation solvents using a database. *Can. J. Chem. Eng.* 63(6), 919–927.

Koyama, K., et al. (1982), The rejection of polar organic solutes in aqueous solution by an interpolymer anionic composite reverse osmosis membrane. *J. Appl. Poly. Sci.* 27, 2845–2855.

Kraetz, L. (1988), Dehydration of alcohol fuels by prevaporation. *Desalination 70*, 481–485.

Krumphanzl, V., and J. Dye (1962), Continuous fermentation and isolation of lactic acid. In *Second Symposium on Continuous Cultivation of Microorganisms* (I. Malek, ed.), Academic Press, New York, pp. 235–244 (published 1964).

Kujawski, W., et al. (1988), Pervaporation of water-alcohol mixtures through Nafion 117 and poly(ethylene-co-styrene sulfonate) membranes. In *Proceedings of Third International Conference on Pervaporation Processes in the Chemical Industry, Nancy, France, September 19–22, 1988* (R. Bakish, ed.), Bakish Materials Corporation, Englewood, NJ, pp. 355–363.

Kurokawa, Y., et al. (1985), Reverse osmosis separation of several alcohols from aqueous solution. *Chem. Eng. Commun. 36*, 333–341.

Larrayoz, M. A., and L. Puigjaner (1987), Study of butanol extraction through pervaporation in acetobutylic fermentation. *Biotechnol. Bioeng. 30*(5): 692–696 (October 5).

Lee, E. K. L., et al. (1983), *Counter-Current Reverse Osmosis for Ethanol-Water Separation*, DOE Report No. DOE/ID/12320-T1, NTIS Order No. DE83008725.

Lee, E. K. L., et al. (1984), Countercurrent reverse osmosis for ethanol-water separation. *American Chemical Society, Division of Polymeric Materials: Science and Engineering*, Vol. 50, pp. 251–255.

Lee, F., and R. H. Pahl (1985a), Solvent screening study and conceptual extractive distillation process to produce anhydrous ethanol from fermentation broth. *Indus. Eng. Chem. Process Des. Dev.* 24(1), 168–172.

Lee, F., and R. H. Pahl (1985b), Use of gasoline to extract ethanol from aqueous solution for producing gasohol. *Indus. Eng. Chem. Process Des. Dev.* 24(2), 250–256.

Lee, Y. M., et al. (1987), Selective organic transport through polyvinylidene fluoride (PVDF) for pervaporation. In *Proceedings of Second International Conference on Pervaporation Processes in the Chemical Industry, San Antonio, TX, March 8–11, 1987* (R. Bakish, ed.), Bakish Materials Corporation, Englewood, NJ, pp. 58–70.

Lee, Y. T., et al. (1988), Pervaporation of water/alcohol mixtures with chemically modified porous glass membranes. *Maku (Membrane) 13*(3), 171–176 (in Japanese with English abstract).

Lee, Y. M., et al. (1989), Sorption, diffusion, and pervaporation of organics in polymer membranes. *J. Membrane Sci. 44*(2/3), 161–181.

Lee, Y. K., et al. (1990), Cationic/anionic interpenetrating polymer network membranes for the pervaporation of ethanol-water mixture. *J. Membrane Sci. 52*, 157–172.

Leeper, S. A., and P. C. Wankat (1982), Gasohol production by extraction of ethanol from water using gasoline as solvent. *Indust. Eng. Chem., Process Des. Dev. 21*(2), 331–334.

Leeper, S. A., et al. (1984), *Membrane Technology and Applications: An Assessment*, DOE Report No. EGG-2282, NTIS Order No. DE84009000.

Leeper, S. A. (1986), Membrane separations in the production of alcohol fuels by fermentation. In *Membrane Separations in Biotechnology* (W. C. McGregor, ed.), Marcel Dekker, New York, pp. 161–200.

Leeper, S. A., and G. T. Tsao (1987), Membrane separations in ethanol recovery: An analysis of two applications of hyperfiltration. *J. Membrane Sci. 30*, 289–312.

Leeper, S. A., and G. F. Andrews (1991), A critical review and evaluation of bioproduction of organic chemicals. *Appl. Biochem. Biotechnol. 28/29*, 499–511.

Leeper, S. A., et al. (1991), *Production of Organic Chemicals via Bioconversion: A Review of the Potential*, DOE Report No. EGG-2645.

Linden, J. C., et al. (1985), Acetone and butanol. In *Comprehensive Biotechnology, Vol. 3, The Practice of Biotechnology: Current Commodity Products* (M. Moo-Young et al., eds.), Pergamon Press, New York, pp. 915–931.

Loeb, S., and M. R. Bloch (1973), Countercurrent flow osmotic process for the production of solutions having a high osmotic pressure. *Desalination 13*, 207.

Lynd, L. R. (1989), Production of ethanol for lignocellulosic materials using thermophilic bacteria: Criteria evaluation of potential and review. *Adv. Biochem. Eng./Biotechnol. 38*, 2–52.

Lynd, L. R. (1990), Large-scale fuel ethanol from lignocellulose: Potential, economics, and research priorities, *Appl. Biochem. Biotechnol. 24/25*, 695–719 (Spring/Summer).

Maiorella, B. L. (1985), Ethanol. In *Comprehensive Biotechnology, Vol. 3, The Practice of Biotechnology: Current Commodity Products* (M. Moo-Young, et al., eds.), Pergamon Press, New York, pp. 861–914.

Maisch, W. F., et al. (1979), Distilled beverages. In *Fermentation Technology (Microbial Technology II)* (H. J. Peppler and D. Perlman, eds.), Academic Press, New York, pp. 79–94.

Masuda, T., et al. (1986), Ethanol-water separation by pervaporation through substituted-polyacetylene membranes. *Polym. J. 18*(7), 565–567.

Masuda, T., et al. (1990), Pervaporation of organic-liquid mixtures through substituted polyacetylene membranes. *J. Membrane Sci. 49*(1), 69–83 (March).

Masuoka, T., et al. (1988), Ethanol-permselective membranes prepared by plasma polymerization. In *Proceedings of Third International Conference on Pervaporation Processes in the Chemical Industry, Nancy, France, September 19–22, 1988* (R. Bakish, ed.), Bakish Materials Corporation, Englewood, NJ, pp. 143–149.

Matsumura, M., and H. Markl (1984), Elimination of ethanol inhibition by perstraction. In *Proceedings of Third European Congress on Biotechnology*.

Matsumura, M., and H. Kataoka (1987), Separation of dilute aqueous butanol and acetone solutions by pervaporation through liquid membranes. *Biotechnol. Bioeng. 30*(7), 887–895 (November).

Matsuura, T., et al. (1977), Predictability of reverse osmosis separation of higher alcohols in dilute aqueous solutions using porous cellulose acetate membranes. *Indus. Eng. Chem. Process Des. Dev. 16*(1), 82–89.

McNeil, B., and B. Kristiansen (1986), The acetone butanol fermentation. *Adv. Appl. Microbiol. 31*, 61–92.

Meares, P. (1988), The sorption and diffusion of vapours in polymers. In *Proceedings of Third International Conference on Pervaporation Processes in the Chemical Industry, Nancy, France, September 19–22, 1988* (R. Bakish, ed.), Bakish Materials Corporation, Englewood, NJ, pp. 12–20.

Mehta, G. D. (1982), Comparison of membrane processes with distillation for alcohol/water separation. *J. Membrane Sci. 12*(1), 1–26.

Mehta, G. D., and M. D. Fraser (1985), A novel extraction process for separating ethanol and water. *Indust. Eng. Chem. Process Des. Dev. 24*(3), 556–560.

Michaels, A. S., and S. L. Matson (1985), Membranes in biotechnology: State of the art. *Desalination 53*, 231–258 (September).

Milsom, P. E., and J. L. Meers (1985), Citric acid. In *Comprehensive Biotechnology, Volume 3: The Practice of Biotechnology: Current Commodity Products* (M. Moo-Young et al., eds.), Pergamon Press, New York, pp. 665–680.

Mohr, C. M., et al. (1988), *Membrane Applications and Research in Food Processing: An Assessment*, DOE Report No. DOE/ID-10210, NTIS Order No. DE88016250.

Mori, Y., and T. Inaba (1990), Ethanol production from starch in a pervaporation membrane bioreactor using *Clostridium thermohydrosulfuricum*. *Biotechnol. Bioeng. 36*, 849–853.

Morigami, Y., et al. (1987), Zero loss solvent recovery. In *Proceedings of Second International Conference on Pervaporation Process in the Chemical Industry, San Antonio, TX, March 8–11, 1987* (R. Bakish, ed.), Bakish Materials Corporation, Englewood, NJ, pp. 200–208.

Morin, M. J., and E. V. Thompson (1988), Pervaporation of alcohol/water solutions through block copolymer membranes. In *Proceedings of Third International Conference on Pervaporation Processes in the Chemical Industry, Nancy, France, September 19–22, 1988* (R. Bakish, ed.), Bakish Materials Corporation, Englewood, NJ, pp. 349–354.

Murakami, H., and N. Igarashi (1981), PBIL tubular reverse osmosis. Application as low-energy concentrators. *Indus. Eng. Chem. Product Res. Dev. 20*(3), 501–508.

Nagase, Y., et al. (1987), Synthesis of substituted-polyacetylene/polydimethylsiloxane graft copolymers and the application for gas separation and pervaporation. In *Proceedings of the 1987 International Congress on Membranes and Membrane Processes*, Tokyo, Japan, Japanese Membrane Society, pp. 558–559.

Nagy, E., et al. (1980), Membrane permeation of water-alcohol mixtures. *J. Membrane Sci. 7*, 109.

Nagy, E., et al. (1983), Pervaporation of alcohol-water mixtures on cellulose hydrate membranes. *J. Membrane Sci. 16*, 79–89.

Nakamura, M., et al. (1988), Separation with fluorinated polymer membranes. *J. Membrane Sci. 36*, 343–351.

Nakao, S., et al. (1987), Continuous ethanol extraction by pervaporation from a membrane bioreactor. *J. Membrane Sci. 30*, 273–287.

Neel, J. (1987), Separation of water-organic liquid mixtures by pervaporation. An insight into the mechanism of the process. In *Proceedings of Second International Conference on Pervaporation Processes in the Chemical Industry, San Antonio, TX, March 8–11, 1987* (R. Bakish, ed.), Bakish Materials Corporation, Englewood, NJ, pp. 35–48.

Niemoller, A., et al. (1988), Radiation-grafted membranes for pervaporation of ethanol/water mixtures. *J. Membrane Sci. 36*, 385–404.

Nguyen, Q. T., et al. (1985), Preparation of membranes from polyacrylonitrile-polyvinyl-pyrrolidone blends and the study of their behavior in the pervaporation of water-organic liquid mixtures. *J. Membrane Sci. 22* (2/3), 245–255.

Nobrega, R., et al. (1988), Separation of ethanol/water mixtures by pervaporation through polyvinyl alcohol membranes. In *Proceedings of Third International Conference on Pervaporation Processes in the Chemical Industry, Nancy, France, September 19–22, 1988* (R. Bakish, ed.), Bakish Materials Corporation, Englewood, NJ, pp. 326–337.

Nomura, Y., et al. (1987), Lactic acid production by electrodialysis fermentation using immobilized growing cells. *Biotechnol. Bioeng. 30*(6), 788–793 (October 20).

Nomura, Y., et al. (1988), Acetic acid production by an electrodialysis fermentation method with a computerized control system. *Appl. Env. Microbiol. 54*(1), 137–142.

Nomura, Y., et al. (1989), Continuous production of acetic acid by electrodialysis bioprocess with a computerized control of fed batch culture, *J. Biotechnol. 12*, 317–326.

Ohya, H., et al. (1981), Concentration of aqueous ethyl alcohol solution by reverse osmosis. *Kagaku Kogaku Ronbunshu 7*, 372.

Ohya, H., et al. (1982), Reverse osmosis concentration of aqueous ethyl alcohol solutions: Analysis of data obtained with composite membranes (PEC). *Kagaku Kogaku Ronbunshu 8*, 144.

Ohya, H. (1983a), Replaceability of distillation process by reverse osmosis: Concentration of dilute aqueous ethyl alcohol solutions *Kagaku Kogaku Ronbunshu 9*, 154–158 (in Japanese with English abstract).

Ohya, H., et al. (1983b), Reverse osmosis concentration of aqueous isopropyl alcohol solutions. *Kagaku Kogaku Ronbunshu 9*, 283–288 (in Japanese with English abstract).

Ohya, H., et al. (1988), Transport of mixed vapors in membrane distillation. In *Proceedings of Third International Conference on Pervaporation Processes in the Chemical Industry, Nancy, France, September 19–22, 1988* (R. Bakish, ed.), Bakish Materials Corporation, Englewood, NJ, pp. 501–517.

Okada, T., and T. Matsuura (1988), A study of the pervaporation of ethyl alcohol/heptane mixtures through porous cellulose acetate membranes. In *Proceedings of Third International Conference on Pervaporation Processes in the Chemical Industry, Nancy, France, September 19–22, 1988* (R. Bakish, ed.), Bakish Materials Corporation, Englewood, NJ, pp. 224–230.

Okamoto, K., et al. (1987), Pervaporation of water-ethanol mixtures through poly-dimethylsiloxane block-copolymer membranes. *Polym. J. 19*(6), 747–756.

Parisi, F. (1989), Advances in lignocellulosics hydrolysis and in the utilization of the hydrolyzates. *Adv. Biochem. Eng. Biotechnol. 38*, 53–87.

Pearce, G. K. (1988), BP's new pervaporation technology. In *Proceedings of Third*

International Conference on Pervaporation Processes in the Chemical Industry, Nancy, France, September 19–22, 1988 (R. Bakish, ed.), Bakish Materials Corporation, Englewood, NJ, pp. 493–500.

Perry, J. H., et al. (eds.) (1963), *Chemical Engineers' Handbook*, 4th Ed., McGraw-Hill, New York, p. 13-6.

Petersen, R. J. (1986), Membranes for desalination. In *Synthetic Membranes* (M. B. Chenowith, ed.), Harwood, Chur, Switzerland, pp. 129–154.

Pham, C. B., et al. (1989), Simultaneous ethanol fermentation and stripping process coupled with rectification. *J. Ferm. Bioeng. 68*(1), 25–31.

Ping, Z. H., et al. (1990), Pervaporation of water-ethanol mixtures through a poly(acrylic acid) grafted polyethylene membrane. Influence of temperature and nature of counterions. *J. Membrane Sci. 48*(2/3), 297–308 (February).

Playne, M. J. (1985), Propionic and butyric acids. In *Comprehensive Biotechnology, Vol. 3, The Practice of Biotechnology: Current Commodity Products* (M. Moo-Young, et al., eds.), Pergamon Press, New York, pp. 731–759.

Prasad, R., and K. K. Sirakar (1990), Hollow fiber solvent extraction: Performances and design. *J. Membrane Sci. 50*, 153–175.

Qin, R., et al. (1990), Separating acetic acid from liquids. *J. Membrane Sci. 50*, 51–55.

Radovanovic, P., et al. (1988), Modeling the pervaporation of ethanol-water mixtures through silicone rubber membrane. In *Proceedings of Third International Conference on pervaporation Processes in the Chemical Industry, Nancy, France, September 19–22, 1988* (R. Bakish, ed.), Bakish Materials Corporation, Englewood, NJ, pp. 181–187.

Radovanovic, P., et al. (1990), Transport of ethanol-water dimers in pervaporation through a silicone rubber membrane. *J. Membrane Sci. 48*(1), 55–65 (January).

Rapin, J. L. (1988), The Betheniville pervaporation unit: The first large-scale productive plant for the dehydration of ethanol. In *Proceedings of Third International Conference on Pervaporation Processes in the Chemical Industry, Nancy, France, September 19–22, 1988* (R. Bakish, ed.), Bakish Materials Corporation, Englewood, NJ, pp. 364–378.

Rautenbach, R., and R. Albrecht (1985a), The separation potential of pervaporation. 1. Discussion or transport equations and comparison with reverse osmosis. *J. Membrane Sci. 25*(1), 1–23.

Rautenbach, R., and R. Albrecht (1985b), The separation potential of pervaporation. 2. Process design and economics. *J. Membrane Sci. 25*(1), 25–54.

Rautenbach, R., and R. Albrecht (1987), Pervaporation and gas permeation: Fundamentals of process design. *Int. Chem. Eng. 27*(1), 10–24.

Rautenbach, R., and R. Albrecht (1989), *Membrane Processes*, John Wiley and Sons, New York.

Reineke, C. E., et al. (1987), Highly water selective cellulosic polyelectrolyte membranes for the pervaporation of alcohol-water mixtures. *J. Membrane Sci. 32*(2/3), 207–221.

Rogers, C. E., and A. Sfirakis (1986), Transport through membranes. In *Synthetic Membranes* (M. B. Chenowith, ed.), Harwood, Chur, Switzerland, pp. 39–51.

Rosenau, J. R., et al. (1986), Lactic acid production from whey via electrodialysis. In *Food Engineering and Process Applications, Vol. 2, Unit Operations* (M. le Maguer and P. J. Jelen, eds.), Elsevier, Amsterdam, pp. 245–250.

Sander, U., and P. Soukup (1988a), Design and operation of a pervaporation plant for ethanol dehydration. *J. Membrane Sci. 36*, 463–475.

Sander, U., and P. Soukup (1988b), Practical experience with pervaporation systems for liquid and vapor separation. In *Proceedings of Third International Conference on Pervaporation Processes in the Chemical Industry, Nancy, France, September 19–22, 1988* (R. Bakish, ed.), Bakish Materials Corporation, Englewood, NJ, pp. 508–518.

Schlicher, L. R., and M. Cheryan (1990), Reverse osmosis of lactic acid fermentation broths. *J. Chem. Technol. Biotechnol. 49*, 129–140.

Schneider, H., et al. (1986), Recent progress in obtaining ethanol from xylose. In *Biotechnology and Renewable Energy* (M. Moo-Young et al., eds.), Elsevier, New York, pp. 161–166.

Schneider, W. H. (1987), Purification of anhydrous organic mixtures by pervaporation. In *Proceedings of Second International Conference on Pervaporation Processes in the Chemical Industry, San Antonio, TX, March 8–11, 1987* (R. Bakish, ed.), Bakish Materials Corporation, Englewood, NJ, pp. 169–175.

Schoutens, G. H., and W. J. Groot (1985), Economic feasibility of the production of iso-propanol-butanol-ethanol fuels from whey permeate. *Process Biochem. 20*(4), 117–121.

Seok, D. R., et al. (1987), Use of pervaporation for separating azeotropic mixtures using two different hollow fiber membranes. *J. Membrane Sci. 33*(1), 71–81.

Serra, A., et al. (1987), A survey of separation systems for fermentation ethanol recovery. *Process Biochem. 22*(5), 154–158.

Sferrazza, R., and C. H. Gooding (1988), An analysis of polybenzimidazole as a pervapora-tion membrane. In *Proceedings of Third International Conference on Pervaporation Processes in the Chemical Industry, Nancy, France, September 19–22, 1988* (R. Bakish, ed.), Bakish Materials Corporation, Englewood, NJ, pp. 97–102.

Shimazu, K., and Matsubara (1987), A solvent screening criterion for multicomponent extractive fermentation. *Chem. Eng. Sci. 42*(3), 499–504.

Shiyao, B., et al. (1988), A study of the pervaporation of isopropyl alcohol/water mixtures by cellulose acetate membranes. In *Proceedings of Third International Conference on Pervaporation Processes in the Chemical Industry, Nancy, France, September 19–22, 1988* (R. Bakish, ed.), Bakish Materials Corporation, Englewood, NJ, pp. 203–211.

Shukla, R., et al. (1989), Acetone-butanol-ethanol (ABE) production in a novel hollow fiber fermentor-extractor. *Biotechnol. Bioeng. 34*, 1158–1166.

Sinegra, J. A., and G. Carta (1987), Sorption of water from alcohol-water mixtures by cation-exchange resins. *Indus. Eng. Chem. Res. 26*(12), 2437–2441.

Smith, B. R., et al. (1977), Separation of lactic acid from lactose fermentation liquors by reverse osmosis. *Aust. J. Dairy Technol. 32*, 23–26.

Sodeck, G., et al. (1987), Application of pervaporation membranes in fermentation processes. In *Proceedings of Second International Conference on Pervaporation Processes in the Chemical Industry, San Antonio, TX, March 8–11, 1987* (R. Bakish, ed.), Bakish Materials Corporation, Englewood, NJ, pp. 157–168.

Sourirajan, S., and T. Matsuura (1985), *Reverse Osmosis/Ultrafiltration Process Principles*, NRCC No. 24188, National Research Council of Canada, Ottawa.

Spitzen, J. W. F., et al. (1987), Solution-diffusion aspects in the separation of ethanol/water mixtures with PVA membranes. In *Proceedings of Second International Conference on pervaporation Processes in the Chemical Industry, San Antonio, TX, March 8–11, 1987* (R. Bakish, ed.), Bakish Materials Corporation, Englewood, NJ, pp. 209–224.

Stelmaszek, J., et al. (1982), Application of different membranes for aqueous mixtures separation. II. Separation of various alcohol-water binary mixtures. *Inz. Chem. Proc. 3*, 205.

Strathmann, H. (1985), Membranes and membrane processes in biotechnology. *Trends Biotechnol. 3*(5), 112–118.

Suematsu, H., et al. (1988), Separation of ethanol-water mixtures by vapor permeation through cellulose acetate. In *Proceedings of Third International Conference on Pervaporation Processes in the Chemical Industry, Nancy, France, September 19–22, 1988* (R. Bakish, ed.), Bakish Materials Corporation, Englewood, NJ, pp. 165–171.

Suzuki, F., et al. (1982), Pervaporation of water-alcohol mixture by cellulose nitrate-poly(methyl acrylate) blended membranes. *Sen-I Gakkaishi 38*(7), T296–T303 (in Japanese with English abstract).

Tanagaki, M., et al. (1987), Selective separation of alcohol from aqueous solution through polymer membrane. In *Proceedings of Second International Conference on Pervaporation Processes in the Chemical Industry, San Antonio, TX, March 8–11, 1987* (R. Bakish, ed.), Bakish Materials Corporation, Englewood, NJ, pp. 126–140.

Tanimura, S., et al. (1990), Ethanol-selective membrane for reverse osmosis of ethanol-water mixture, *AIChE J. 36*(11), 1118–1120 (July).

Tegtmeier, U. (1985), Process design for energy saving ethanol production. *Biotechnol. Lett. 7*(2), 129–134.

teHennepe, H. J. C., et al. (1987), Pervaporation with zeolite filled silicone rubber membranes. In *Proceedings of Second International Conference on Pervaporation Processes in the Chemical Industry, San Antonio, TX, March 8–11, 1987* (R. Bakish, ed.), Bakish Materials Corporation, Englewood, NJ, pp. 71–78.

Terada, I., et al. (1988), Water/ethanol permeation properties through poly(hydroxy-methylene) and poly(hydroxymethylene-co-fluoroolefin) membrane by pervaporation method. *Desalination 70*, 455–463.

Uragami, T., et al. (1988), Permeation and separation characteristics of aqueous alcohol solutions through hydrophilic and hydrophobic membranes by pervaporation and evapomeation. In *Proceedings of Third International Conference on Pervaporation Processes in the Chemical Industry, Nancy, France, September 19–22, 1988* (R. Bakish, ed.), Bakish Materials Corporation, Englewood, NJ, pp. 127–133.

Vickroy, T. B. (1985), Lactic acid. In *Comprehensive Biotechnology, Vol. 3, The Practice of Biotechnology: Current Commodity Products* (M. Moo-Young, et al., eds.), Pergamon Press, New York, pp. 761–776.

Vijaikishore, P., and N. G. Karanth (1984), Glycerol production by fermentation. *Appl. Biochem. Biotechnol. 9*, 243–253.

Vijaikishore, P., et al. (1987), Concentration of glycerol in fermentation broths by reverse osmosis. *J. Microbiol. Biotechnol. 2*(1), 22–27.

Walton, M. T., and J. L. Martin (1979), Production of butanol-acetone by fermentation. In *Microbial Processes* (Fermentation Technology II) (H. J. Peppler and D. Perlman, eds.), Academic Press, New York, pp. 188–209.

Watson, J. M., and P. A. Payne (1990), A study of organic compound pervaporation through silicone rubber. *J. Membrane Sci.* *49*(2), 171–205 (April 1).

Wecker, M. S. A., and R. R. Zall (1987a), Fermentation strategies: Acetaldehyde or ethanol? *Process Biochem.* *22*(5), 135–138.

Wecker, M. S. A., and R. R. Zall (1987b), Production of acetaldehyde by *Zymomonas mobilis*. *Appl. Env. Microbiol.* *53*(12), 2815–2820.

Wenzlaff, A., et al. (1985), Pervaporation of water-ethanol through ion exchange membranes. *J. Membrane Sci.* *22*(2/3), 333–344.

Wesslein, M., et al. (1988), Pervaporation of binary and multicomponent mixtures using PVA-membranes: Experiments and model calculations. In *Proceedings of Third International Conference on Pervaporation Processes in the Chemical Industry, Nancy, France, September 19–22, 1988*) (R. Bakish, ed.), Bakish Materials Corporation, Englewood, NJ, pp. 172–180.

Wesslein, M., et al. (1990), Pervaporation of liquid mixtures through poly(vinyl alcohol) (PVA) membranes. I. Study of water containing binary systems with complete and partial miscibility. *J. Membrane Sci.* *51*, 169–179.

Yamada, S., and T. Nakagawa (1987), Separation of alcohol-water solution with copolymer membranes. In *Proceedings of Second International Conference on Pervaporation Processes in the Chemical Industry, San Antonio, TX, March 8–11, 1987* (R. Bakish, ed.), Bakish Materials Corporation, Englewood, NJ, pp. 176–185.

Yoshikawa, M., et al. (1984), Pervaporation of water-ethanol mixture through poly(male-imide-co-acrylonitrile) membrane. *J. Membrane Sci.* *22*(1), 125–127.

Yoshikawa, M., et al. (1987), Selective separation of aqueous ethanol solution through synthetic polymer membranes. In *Proceedings of 1987 International Congress on Membranes and Membrane Processes*, Tokyo, Japan, Japanese Membrane Society, pp. 594–595.

Young, J. K., et al. (1989), Bioprocessing of commodity chemicals: Research directions for government, industry, and academic. *Appl. Biochem. Biotechnol.* *20/21*, 339–355.

Zacchi, G., and A. Axelsson (1989), Economic evaluation of preconcentration in production of ethanol from dilute sugar solutions. *Biotechnol. Bioeng.* *34*, 223–233.

6

Commercial Applications of Emulsion Liquid Membranes

Robert P. Cahn *Private Consultant, Millburn, New Jersey*

Norman N. Li *Research and Technology, Allied-Signal, Inc., Des Plaines, Illinois*

INTRODUCTION

The emulsion liquid membrane (ELM) concept constitutes an elegant way of interposing an extensive yet extremely thin membrane between two liquid phases without any mechanical appurtenances. Its basic simplicity allows easy regeneration, freeing the user of many of the potential problems inherent in conventional membrane processes. However, since the membrane is a liquid, ELM does not permit pressure-driven separations and is thus restricted to concentration-driven permeation processes.

While many potential applications of ELM have been investigated, especially by Exxon Research in the 1970s [1–7], commercial uses of the concept have only recently been realized [8–22]. These uses exploit two opposite properties of the liquid membrane: (a) its ability to act as a selective transfer agent for at least one of the constituents of the phases being kept apart by the membrane, and (b) its capacity to maintain its integrity and thus keep separate for an extended time two phases containing reagents but, by imposing appropriate conditions, destroy this integrity, thus allowing the two phases to interact. Liquid membrane emulsions, being formed as a result of surface-active forces, are only inherently stable under certain limited conditions of temperature, shear, loading, acidity, composition,

195

etc. Separation applications described in (a) must operate below the critical levels of these conditions, while the reaction applications described in (b) operate first in the stable and then in the unstable regimes, i.e., both below and above these critical levels.

CURRENT EMULSION LIQUID MEMBRANE STATUS

Considerable research, principally in university laboratories, has ben carried out on the ELM concept since its original invention by Li [1]. An excellent review of the development of the theory covering mass transfer of the solute into the emulsion globules is presented by Ho [12]. Both a spherical shell approach and a model using an advancing front diffusion into the emulsion globules has been used to allow design of ELM equipment. The spherical shell approach, which assumes that diffusion is controlled by a thin spherical shell of membrane phase separating the internal and external phases, is only approximate, and allows rough sizing and qualitative comparison of different systems. The advancing front or emulsion globule model is much more rigorous and complex, but permits optimization of the many variables controlling ELM.

The bulk of the work into ELM applications has been carried out in two areas, referred to as type 1 and type 2 facilitation [12]. In type 1 facilitation, the solute is reacted with a material dissolved in the internal phase encapsulated in the emulsion, forming a nonpermeable compound. This effectively reduces the concentration of the solute to zero in the internal phase. Examples of this are membrane-soluble weak acids and bases, such as phenol, ammonia, organic acids and bases, all of which can be efficiently neutralized and thus converted to membrane-insoluble salts, i.e., ions. On the other hand, type 2 facilitation utilizes a complexing agent for the solute which is incorporated into the membrane phase. The solute, generally an ion, is carried across the membrane and into the internal membrane phase driven by counterdiffusing ions, usually H^+ or OH^-. The counterion is generally maintained at a relatively high concentration within the emulsion. Examples of type 2 transport are metallic ions, such as Cu^{2+}, $Cr(VI)^+$, Ni^{2+}, Co^{2+} or Zn^{2+} and their complexes.

Both type 1 and 2 facilitation allow considerable concentration of the solute. Therein lies one of the potential attractions of this process in treating dilute solutions and waste streams, i.e., concentrating the problem into a small, manageable volume.

The ELM emulsion is first prepared in a high-shear mixing device and is then used in the ELM process similar to the solvent in solvent extraction. With proper emulsion formulation regarding surfactants, thickening additives, and complexing agents, emulsions can be prepared which are stable in the extraction process, but allow efficient demulsification of the spent emulsion, usually in an electrostatic coalescer [13,14]. The internal phase droplets in the emulsion are micrometer size (1–3 μm), and membrane thickness is of the order of 1 μm.

SEPARATION APPLICATIONS

Three separations using emulsion liquid membranes have been commercialized:

1. The recovery of zinc from rayon plant effluents
2. The removal of phenol from wastewater at a plastics plant
3. Treatment of cyanide-containing wastewater from gold mining

A commercial zinc recovery facility has been built and operated in Austria. The research leading up to its design and construction was described in a paper presented by Marr et al. [8] at the 1987 Engineering Foundation Conference on Separations. The discussion presented below will draw extensively on this paper. A company in Europe is marketing this application.

Several phenol removal plants have been installed in China. The first commercial facility is described briefly in a news release by the South China Institute of Technology at Guangzhou [9], and more fully in a paper presented at the Second International Conference on Separation Technology in 1987 [10]. It may be interesting to comment from a historical point of view that it was the demonstration of phenol removal from wastewater using a caustic containing liquid membrane emulsion which led the present authors to investigate the broad application of LM in the separation of aqueous systems [2].

Another commercial application of LM, announced in a Chinese publication, *Kexue Bao* (*The Newspaper of Science*) [11], has recently been discussed further at ICOM '90 [16,17]. It involves the treatment of cyanide-containing wastewater from gold mining, reducing the cyanide content to a level which allows reuse of the water in the installation.

Work is continuing to develop existing and to find further commercial applications for the ELM process. Yan et al. [15] report pilot scale work on the removal of Cr(VI) from sodium dichromate plant effluents, successfully reducing the Cr level from 100 ppm to below 0.5 ppm in a two-stage treat. According to Wu [16], gold can be extracted from an acidic feed solution with a caustic-containing emulsion, using a sulfoxide as carrier in the membrane. Extraction of gold from an alkaline cyanide feed is described by Jin [17], who showed that cyanide can also be removed. Pilot plant work on phenol extraction from coking plant wastewater streams is covered by Wang [18]. In this research, an RDC contractor with conical holes was found to be especially effective in improving emulsion-feed contacting.

These separation applications will be discussed below, based strictly on the information contained in the cited publications. Finally, a few comments will be made on possible further commercial applications, and some potential improvements which could affect the scope of ELM applications.

Principles of Liquid Membrane Separation Processes

Liquid membrane processes, as the name implies, are based on the interposition of a membrane of an immiscible liquid between two, usually miscible phases.

Transfer of solute occurs selectively through this membrane from one phase into the other. Two kinds of liquid membranes are being considered at present: (a) supported liquid membranes and (b) emulsion liquid membranes (ELM).

The support for (a) is usually in the form of a porous solid membrane which preferentially soaks up and holds the liquid membrane fluid. While extensive work is underway on this concept, first disclosed in U.S. Patent 3,244,763 [23], no commercial application has as yet been realized.

On the other hand, the emulsion liquid membrane, invented and developed by Li [1,24,25], uses surface forces to maintain a very thin yet stable immiscible membrane between the two miscible phases. Commercial use of this technique was recently made in the chemical and petroleum industries, and this is the subject of the present chapter.

The emulsion consists of a continuous membrane phase and an internal droplet phase, which is usually the recipient of the solute which is allowed to permeate from an external feed phase through the membrane when the emulsion is dispersed into the feed by appropriate agitation. Figure 1 illustrates schematically what a typical emulsion LM system looks like. The operation can be carried out

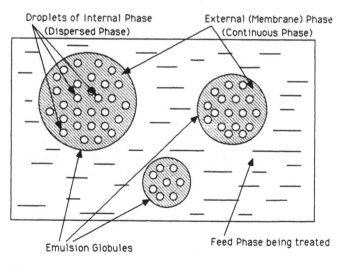

Figure 1   Schematic illustration of emulsion LM system. The LM emulsion is dispersed in the form of emulsion globules into the feed being treated. The emulsion in the globules remains intact. Permeation of substances dissolved in the feed phase can occur through the continuous (membrane) phase of the emulsion into the dispersed droplet phase of the emulsion. Two conditions have to be met for this transfer to occur: (a) the dissolved substance has to be somewhat soluble in the membrane phase and (b) there has to be a driving force, such as a concentration gradient.

batchwise or continuously, single stage or multistage, cocurrently or countercurrently. Laboratory tests are usually done in a batch system on account of the simplicity of the method and the relatively straightforward translatability of batch to continuous data.

When the feed and the internal droplet phases are aqueous, the membrane is generally a hydrocarbon oil containing a surfactant. Frequently, the solute to be removed is only sparingly soluble in the hydrocarbon. In this case, an appropriate solubilizer must be added to the hydrocarbon membrane to enhance this solubility, and consequently the permeability of the solute through the membrane.

Unless a "driving force" is added, permeation of the solute from feed to internal phase of the emulsion is a short-lived phenomenon, since it will stop as soon as the concentrations of the solute on the two sides of the membrane are equal. Addition to the internal phase of a reagent, which consumes or alters the solute upon its arrival inside the droplets, is one way in which a concentration gradient can be maintained. Another very effective method useful for ion transport is to utilize another ion as a "pump." Its counterdiffusion effectively drives the desired ion through the membrane which is laced with a selective ion exchange compound. The equilibrium between this ion exchange material, the desired ion and the counterions, usually protons, establishes the flow of the desired ion into the internal phase, even under conditions where there is a sizable reverse concentration gradient.

Rate equations governing the permeation of a solute into the LM emulsion are based on the conventional membrane approach, which assumes Fick's law to be controlling:

$$\text{Solute flux} = \frac{dN_B}{d\theta} = \frac{D_B k_B \Delta C_B A}{\Delta x} \tag{1}$$

where

$dN_B/d\theta$ is moles of solute B diffusing into the emulsion/unit time θ

D_B is the diffusion coefficient of B through the membrane

k_B is the distribution coefficient of B between the aqueous and the membrane phases

ΔC_B is the concentration difference of B between the feed and the internal aqueous phase

A is the membrane surface area, and

Δx is the membrane thickness

For a batch LM system, if we assume that the effective concentration of B in the internal phase is zero, then

$$\frac{dN_B}{d\theta} = b'\left(\frac{-dC_{Bf}}{d\theta}\right) \tag{2}$$

and

$$A = b'' \left(\frac{V_e}{V_w} \right)$$

(3)

where

b', b'' are proportionality constants

C_{Bf} is the concentration of B in the feed phase, and

V_e/V_w is the treat ratio, volume emulsion/volume aqueous feed in the contactor

Then by lumping the proportionality constants b', b'', and the parameters D_B and k_B into a single permeation constant D', we cast the permeation rate equation for LM extraction into the form:

$$\frac{dC_{Bf}}{d\theta} = -D' \left[C_{Bf} \left(\frac{V_e}{V_w} \right) \right]$$

(4)

Equation 4, which can be easily solved for laboratory batch extraction, has been used extensively to establish extraction rates and to evaluate the effect of variables on these rates.

Following is a partial list of the effect of different parameters on the liquid membrane permeation rate constant D':

1. D' increases with increasing solubility of B in the membrane phase.
2. When the solubility of B in the membrane phase is dependent on a complexing or ion exchange material dissolved in that phase, D' will increase as the concentration of the solubilizing agent is increased, but will level off at some point.
3. D' increases with decreasing viscosity of the membrane phase.
4. D' increases up to a point as the ratio of internal to membrane phases in the emulsion is increased.
5. Increased agitation improves D' until the integrity of the emulsion is jeopardized.
6. Better emulsification, i.e., smaller internal phase droplets, improves D'.

After this relatively cursory review of the general principles of LM emulsion processes, we will now turn our attention to some systems which, based on literature information, have found commercial use.

Recovery of Zinc

Rayon is spun into a bath of concentrated zinc sulfate solution, so that the washing effluents from these plants are high in zinc. Both environmental considerations and economics necessitate removal and, if possible, recovery of the zinc, preferably as a saturated $ZnSO_4$ solution.

While competing processes, such as precipitation, permit removal, only the advent of the liquid membrane process allowed economic recovery and reuse of the zinc. After two years of pilot plant work, a commercial size LM plant has been constructed on a scale of 75 m³/hr, recently expanded to 350 m³/hr. In the order of 90% of the zinc in the effluent is recovered and recycled.

Since zinc ion is insoluble in the hydrocarbon membrane phase of the LM system, a solubilizer had to be found. After much experimentation, DTPA, or di(2-ethylhexyl)dithiophosphoric acid (Hoechst), turned out to be a satisfactory agent, with a good selectivity of zinc vs. calcium ions. An interesting feature is that although this agent is not popular for solvent extraction on account of its slow stripping kinetics, this property is not a handicap in LM processing where the large interfacial area available for stripping, the internal interfacial area in the emulsion, obviates this problem. The design of the homogenizer and the electrostatic demulsifier had to be optimized for this highly corrosive system. Hastelloy was required for the homogenizer, and the optimum demulsification conditions required 20 kV and much higher than the conventional 50-Hz frequency, namely, as high as 4–10 kHz.

The internal phase started out as 250–300 g/liter H_2SO_4, but the loaded material reached 55–60 g/liter Zn, with the acid concentration dropping to 100 g/liter. Protons were effective as pumping ions, allowing considerable buildup of the zinc concentration in the internal while the feed zinc was reduced from 100–350 mg/liter to a level of the order of 1 mg/liter. Countercurrency was achieved by carrying out the extraction in an agitated column with internals of the Oldshoe–Rushton type, i.e., impellers alternating with baffles to demarcate stages.

Tables 1 and 2 list the major process conditions and apparatus dimensions of

Table 1 Highlights of Zinc Application

Application:	Rayon plant waste	
Developers:	Lenzing AG, Austria	
Plant sizes:	Pilot plant	1000 liters/hr (2 gpm)
	Commercial plant	75 m³/hr (150 gpm)
	planned	350 m³/hr (700 gpm)
Complexing agent:	Di(2-ethylhexyl)dithiophosphoric acid (DTPA)	
Membrane composition:	3% DTPA	
	3% Exxon PX100 surfactant	
	94% Shellsol T (1.85 cp)	
Demulsifier:	20 kV; 50 Hz, 5 kWh/m³ emulsion	
Zn removal:	98%	

Table 2 Zinc E.L.M. Extraction Plants

		Pilot plant	Comm. plant
Feed rate	m³/hr	0.7	70–75 (350)
Zn conc.	mg/liter in	450–500	350
	out	2–10	5
Membr. phase,	liters/hr	40	7000
PX-100	wt. %	2	3
DTPA	wt. %	3	3
Shellsol T	wt. %	95	94
I.R.	liters/hr	4	300
H_2SO_4	g/liter in	250	250–300
	out	100	100
Zn^{2+}	g/liter out	55–65	55–60
Coalescer	kV	20	
	Hz	50	500 (5–10,000)
	kWh/m³	5	
Extractor	Ht., m	7	10
	Diam., mm	150	1600

the pilot facility and the first commercial plant. Initial operation of the commercial plant with a feed of higher density yet lower zinc concentration than the design feed indicated a problem. The lower residence time of the emulsion droplets in the column on account of the increased density differential (feed was the continuous, heavier phase) reduced zinc cleanup to the point that design effluent concentration could not be achieved.

This problem was solved by modifying the emulsion formulation by increasing the membrane/internal reagent ratio. At the same internal/feed ratio, the emulsion/feed ratio was thereby increased, resulting in a longer emulsion residence time in the column. This change did overload the demulsifier, however, but this problem was addressed by optimizing the frequency of the applied voltage.

There may be, incidentally, another way of modifying the residence time of two phases flowing through a continuous, stirred contacting zone. This method is described by Hatton et al. [26] for LM systems and is similar to a technique developed for solvent extraction by Davy McKee [27,28]. As shown in Figure 2, it utilizes partial settling of the phases in a zone separate from the mixing zone allowing withdrawal of the two countercurrent phases in a ratio far different from the ratio as maintained in the mixing zone. This technique was specifically developed for LM extraction, since it was felt that the low emulsion/feed ratios encountered in LM processes could easily lead to unsatisfactory contacting in the mixing stage, precisely as experienced in the Austrian plant.

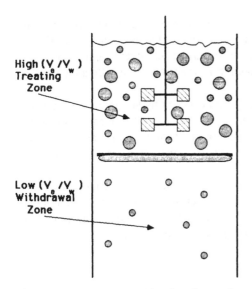

High (V$_e$/V$_w$)
Treating
Zone

Low (V$_e$/V$_w$)
Withdrawal
Zone

Figure 2 High treat ratio but low flow ratio.

The direct operating cost breakdown, as presented in the Marr et al. paper [8], and shown here in Table 3, indicates that evaporation of the product internal phase from about 60 g/liter Zn to a saturated zinc sulfate solution (above 130 g/liter Zn) is responsible for about a quarter of the total operating cost. It may very well be that in place of evaporation, membrane permeation of water from the dilute internal phase product into concentrated makeup sulfuric acid may be an economical way to remove water from the internal solution. Solid rather than liquid membranes

Table 3 Economics of Zinc
Removal by ELM[a]

Factor	$/kg Zn
Sulfuric acid	0.15
Neutralization	0.14
Power	0.06
Organic losses	0.08
Evaporation	0.12
Total	0.54

[a]Direct costs, based on 1987 value
of $0.067/shilling.

may be preferred for this operation since water diffusion, a phenomenon encountered in liquid membrane extraction and usually referred to as "osmotic swell," is not a rapid process. Also, concentrated sulfuric acid, even when somewhat diluted, may be a rather harsh environment for the surfactant usually added to the membrane phase in ELM processes.

Marr et al. [8] investigated the effect of different contacting variables on the rate of osmosis (water dilution of the internal phase) vs. the rate of zinc removal from the feed. They found that increasing the holdup of emulsion phase in the column by increasing the membrane phase was preferable to achieving this increase by more violent agitation.

One way to improve the ratio of zinc/water transfer into the internal phase is to carry out part of the zinc removal by solvent extraction with the recycle membrane phase prior to using it to prepare the LM emulsion. This technique is described in U.S. Patent 4,086,163 [29] and illustrated schematically in Figure 3. Since zinc

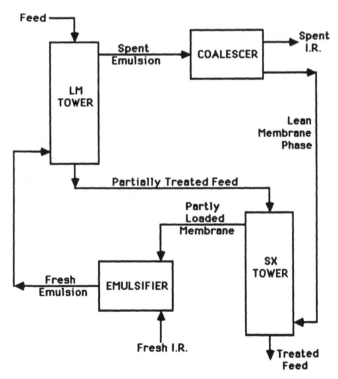

Figure 3   Combined LM/SX to reduce swell. Osmotic "swell" of the internal reagent can be reduced substantially if part of the extraction duty is done by the coalesced membrane phase in a liquid/liquid extractor prior to feeding this phase to the emulsifier to make fresh LM emulsion. There is no osmosis of water in the extractor, since the internal reagent is not present.

removal by solvent extraction is not debited with any concomitant water transfer, and the process demand on the LM step is reduced by the extent of zinc removal achieved in the solvent extraction, the water transfer, i.e., overall osmotic swell, in this combined SX/LM operation is less than if the zinc removal is carried out by LM extraction alone.

In summary, it is very gratifying to hear of a successful commercialization of the liquid membrane separation process, realized after extensive pilot plant experiments as well as subsequent commercial plant optimizations.

Phenol Removal

The removal of phenol from the wastewater of a plastics plant is described in a paper by the process developers, the South China Institute of Technology (SCIT) [9]. Based on the paper by Professor Zhang et al., presented at the Schloss Elmau Separations Conference in Germany in 1987 [10], the pilot plant feed rate was of the order of 200–250 liters/hr, the commercial plants were about twice that size. Using a 1:1 emulsion of hydrocarbon/5 wt % aqueous NaOH solution, the pilot plant removed 99.96% of the 350–925 mg/liter of phenol in the feed, with the effluent consistently containing less than 0.5 mg/liter phenol and 2.27 ml/liter of residual oil.

A special surfactant and an electric coalescer for this process were developed by SCIT. A colloid mill was used for emulsion preparation, and the extraction took place in a two-stage rotary disk column. The membrane formulation contained 3.5 wt % surfactant, 6.7 wt % liquid paraffin, with kerosene as the solvent. The emulsion proved to be very stable; circulation for 4 hr in the column indicated no osmotic swell or intolerable leakage. Demulsification was carried out at a rate of 32–55 liters/hr at 20 kV, 0.6–25 kHz, without any evidence of sparking or "spongy emulsion" loss. The recovered oil phase, containing about 1% aqueous phase, can be reused directly, and 24 recycles without surfactant makeup have been demonstrated. In April of 1987, the commercial demonstration plant was reported to have operated successfully for almost a year. Some pertinent information regarding the demonstration plant and typical operating data are shown in Table 4. This operation also proved that the membrane phase could be recycled repeatedly without detrimental effect on the extraction efficiency of the unit.

The economics of the process, as reported by the SCIT researchers, are shown in Table 5. They compare favorably with competitive cleanup techniques. Based on the successful demonstration of the process and the favorable economics, the investigators report that "recently ten factories have asked to adopt this process in treating their wastes."

Cyanide Removal

In an October 1987 announcement the Chinese publication *Kexue Bao* reported another commercial application of the emulsion liquid membrane technology [11].

Table 4 Phenol Plant Performance

Emulsion:	Membrane phase: 1 vol.	
		wt %
	Liq. paraffin (d = 0.84 g/cm³)	6.7
	Surfactant "LMS-2"	3.5
	Kerosene (d = 0.79 g/cm³)	89.8
	I.R.: 1 vol.	
	NaOH	5.0

Treating columns: 2 rotary disk columns

	Each: Internal diam.	22 cm
	Height	3.5 m
	Stages/column	2

Coalescer: 20 kV; 50 mA max; 0.6–25 kHz, 300 VA; 20-liter capacity; Insulated horizontal electrodes

	Phenol conc. (mg/liter)		
Oil phase reuse no.	Inlet	Outlet	Phenol removal (%)
0	916.5	0.29–0.37	99.96
10	1410	0.08–0.32	99.98
23	999.6	0.04–0.09	99.99

No measurable oil entrainment; no osmotic swell; phenol in I.R. as high as 53.6 g/liter.

Table 5 Economics of Phenol Removal[a]

Direct cost	Unit/ton	Unit cost ($)	Cost/ton ($)
Oil phase	0.225 liter		
Kerosene	0.161 kg	0.19	0.03
Liq. Paraff.	0.012 kg	0.73	0.01
LMS-2	0.0062 kg	5.13	0.03
I.R.	16.51 liter		
NaOH (s)	0.825 kg	0.22	0.18
Electricity	3 kWh	0.04	0.12
Total			0.37

[a]Direct costs, $/ton of 1000 mg/liter wastewater, based on $0.27/yuan. Investment for a 4-ton/8-hr plant: $19,000.

Moderately sized gold mines in China have a problem with disposal of their cyanide-containing waste liquors as well as gold losses in these effluents. The Chemical Physics Institute of the Chinese Academy of Science in Dalien in collaboration with the gold mine in Chashe city has successfully developed the necessary technology to allow treatment of the cyanide-containing wastewater with LM.

According to the announcement, the Chemical Physics Institute solved a number of technical problems such as membrane composition, selection of surfactants and extractants, and determining the optimum conditions for de-mulsification. Plant tests have shown that cyanide concentration was lowered to 0.5%, which not only satisfied effluent release standards, but also permitted reuse of the treated water. This has led to increased productivity in the gold mine. The process is simple, and is claimed to involve low capital investment and operating cost. A recent paper by Jin [17] gave more details about this application. Although there may be many extractants to be used, an amine or similar anion exchange resin, like Alamine [30], with an internal phase of caustic solution could consti-tute an effective system.

SEGREGATION OF REAGENTS BY ENCAPSULATION

In oil drilling, costly delays can occur when drilling fluid, which is used to carry rock chips to the surface and to control well pressure, is lost by leakage into underground formations. At a cost of $7000 per day for drilling on land, and up to $100,000 daily for offshore operation, stopping the drilling to correct this leakage problem can be an expensive interruption.

Conventional methods to correct the loss of drilling fluid involve either plugging the leak by a variety of materials or isolating the loss zone by placing an extra steel casing over it, often at considerable cost. Various plugging techniques, not always effective, include the use of granular material pumped downhole to plug the leak, the use of cement which is time consuming, or relatively unreliable procedures which depend on mixing two or more reagents downhole.

To solve this problem, Exxon developed a unique fluid which can be pumped down the drill pipe to the bottom of the well, where it would then thicken into a paste [19–22]. As shown in Figure 4, the fluid consists of water droplets and granules of water-swellable clay finely dispersed in a continuous oil phase. Each water droplet is surrounded by a tough liquid film containing a water-soluble polymer that strengthens the film. When the slurry is pumped into the well, the tough film is ruptured by high shear forces when the slurry passes through the nozzles in the drill bit. This releases the water, allowing it to react with the clay, so that the shear-thickening fluid emerges from the nozzle with the consistency of a paste. The resultant paste can effectively seal the loss zone without the need for costly delays.

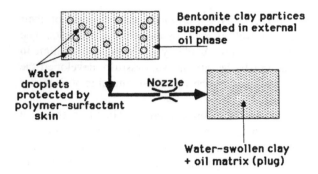

Figure 4 Well control fluid. The fluid consists of a continuous oil phase, in which droplets of water, encased in a tough polymer film, and swellable clay particles are dispersed. This pumpable fluid can be injected into a defective oil well through a high-shear nozzle, breaking the polymer film and allowing interaction between the clay and water. The resultant paste effectively seals the loss zone in the well.

The formation has been successfully field-tested in a number of problem oil wells in the United States that had proven intractable to conventional fluid loss control techniques.

Here is an example of the commercialization of liquid membrane technology, not in the field of application for which it was originally intended, namely separations, but in the field of encapsulation. The shear-thickening fluid capitalizes on reagent emulsion properties developed and the expertise accumulated in the liquid membrane area over nearly a decade of fundamental and applied research.

SOME POSSIBLE FUTURE IMPROVEMENTS

It may be interesting to speculate on some possible future improvements which could have a significant effect on liquid membrane technology and the extent of its commercialization.

There seem to be two major areas where improvements could be made:

1. The high chemicals cost, since the concentrated internal phase is generally consumed in the process, and the extracted material sometimes discarded.
2. The need of breaking the emulsion, and the cost of the associated equipment and power consumption.

The chemicals consumption can potentially be improved substantially when the internal phase is a base or acid, and the driving force reaction is a neutralization. A possible scheme is illustrated in Figure 5. Instead of discarding the spent internal reagent, the neutralized or partially neutralized material can be subjected

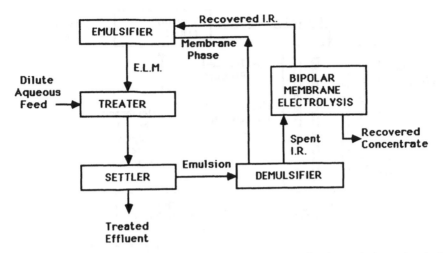

Figure 5 Combined emulsion LM and bipolar membrane. The internal phase chemical reagent, frequently an acid or base, is used up in the treating process. It can be regenerated by sending the spent internal reagent to a bipolar membrane cell, which not only regenerates the original acid or base, but also releases the compound removed from the feed in relatively concentrated form.

to a bipolar membrane process developed by Allied Signal, Inc., which effectively regenerates fresh internal reagent and liberates the permeate in recoverable, unneutralized form [31]. In the case of phenol removal with caustic emulsion, the spent sodium phenate would be regenerated to caustic solution suitable for incorporation into recycle emulsion, and phenol would be generated as such. The economics of such a step vs. the use of fresh makeup chemical and cost of disposal of the discarded internal phase would have to be evaluated. It should be noted that use of the bipolar membrane regeneration technique opens up the potential use of more expensive internal reagents, if such a reagent is preferable for process reasons, since it will be regenerated and recycled without substantial loss.

The cost of demulsification may be reduced if a less stable emulsion could be employed without running the danger of leakage and actual emulsion breakup. One possible way to achieve this goal, as shown in Figure 6, would be to use a porous, solid membrane, such as in the form of hollow fibers, to carry out contacting between feed, on the outside of the fibers, i.e., on the shell side, and the unstable emulsion flowing through the lumen of the fibers. If the fiber material is preferentially wetted by the membrane phase, transfer will take place just as regular LM processing. The scheme is very similar to a supported liquid membrane process with all its advantages and disadvantages. Emulsion stability can be

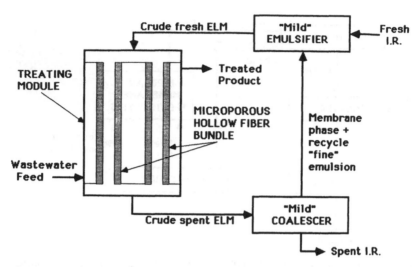

Figure 6 Combined emulsion LM and supported LM. It may be possible to use a less stable emulsion if the feed/emulsion contacting is done in a porous membrane contactor. This would not only reduce the difficulty of breaking the spent emulsion, but may also permit use of the emulsion LM technique to treat feeds containing suspended solids, so-called extraction-in-pulp.

very low, since the integrity of the emulsion is maintained by turbulence as the emulsion flows through the fibers, but leakage of internal into the feed or vice versa is prevented by the porous membrane, which is selective to the membrane phase, barring both feed and internal phase from passing through it. The concept would also allow use of LM extraction for mineral extraction "in pulp," i.e., the extraction of such metal values as copper or uranium aqueous leach liquors which still contain substantial amounts of solids. Attempts to carry out such separations with conventional extraction and LM emulsion techniques result in the rapid formation of sludges and pastes (called "crud" in hydrometallurgy).

In solvent extraction, it has been found that because hollow fibers can be packed into a given extraction volume with a surprisingly high density, the total volume required for this kind of extraction may not be any greater than that required for conventional extraction. A similar result is expected for liquid membrane extraction and demulsification demands can be sharply reduced.

There is a good possibility that other interesting combinations of emulsion liquid membranes with the rapidly developing general membrane technology could lead to commercially viable applications.

REFERENCES

1. N. N. Li, U.S. Patent 3,410,794 (November 12, 1968).
2. N. N. Li, R. P. Cahn, and A. L. Shrier, U.S. Patent 3,617,546 (November 2, 1971).
3. N. N. Li, and A. L. Shrier, *Recent Developments in Separation Science*, Vol. 1, Chemical Rubber Co., Cleveland, 1972, p. 163.
4. R. P. Cahn, and N. N. Li, *Sep. Sci.* 9(6), 505–519 (1974).
5. R. P. Cahn, N. N. Li, and R. R. Minday, *Env. Sci. Technol.* 12, 1051–1055 (1978).
6. N. N. Li, R. P. Cahn, D. Naden, and R. W. M. Lai, *Hydrometallurgy 9*, 277–305 (1983).
7. T. Kitagawa, Y. Nishikawa, J. W. Frankenfeld, and N. N. Li, *Environ. Sci. Technol. 11*, 602–605 (1977).
8. J. Draxler, R. Marr, and M. Protsch, Commercial-Scale Extraction of Zinc by Liquid Emulsion Membranes, Paper presented at Engineering Foundation Conference, Schloss Elmau, Federal Republic of Germany, April 26–May 1, 1987. *Separation Technology* (N. N. Li and H. Strathmann, eds.), A. I. Ch.E., New York, 1988, p. 204.
9. Zhang Xiujuan et al., News Release, An Industrial Application of Liquid Membrane Separation for the Phenol Wastewater Treatment, South China Institute of Technology, Guangzhou, China.
10. Zhang Xiujuan, Liu Jianghong, Fan Qiongjia, Lian Quingtang, Zhang Xingtai, and Lu Tiangsi, Industrial Application of Liquid Membrane Separation for Phenolic Wastewater Treatment, Paper presented at Engineering Foundation Conference, Schloss Elmau, Federal Republic of Germany, April 26–May 1, 1987. *Separation Technology* (N. N. Li and H. Strathmann, eds.), A.I.Ch.E., New York, 1988, p. 190.
11. Treatment of Cyanide-Containing Waste Water from Gold Mine Operation by Liquid Membrane Technology, news release in *Kexue Bao* (*Newspaper of Science*), China, October 16, 1987.
12. W. S. Ho, Emulsion Liquid Membranes: A Review, Paper presented at the 1990 International Congress on Membranes and Membrane Processes, (ICOM'90), Chicago, IL, August 20–24, 1990. See *ICOM '90 Proceedings* AIChE, New York, 1990.
13. E. C. Hsu and N. N. Li, *Sep. Sci. Technol. 20*, 115 (1985).
14. Z. Feng, X. Wang and X. Zhang, *Water Treatment 3*, 320 (1988).
15. Z. Yan, S. Li, Y. Chu, and B. Liu, Treatment of Wastewater Containing High Concentration of Cr(VI) Using Liquid Membrane, *ICOM '90 Proceedings*, AIChE, New York, 1990.
16. W. Wu and C. Chen, *Water Treatment 5*, 277 (1990).
17. M. Jin and Y. Zhang, Study on Extraction of Gold and Cyanide from Alkaline Cyanide Solution by Liquid Membrane, *ICOM '90 Proceedings*, AIChE, New York, 1990.
18. J. Wang, B. Wang, and Z. Cui, *Water Treatment 5*, 348 (1990).
19. Exxon Publication *Lamp*.
20. C. R. Dawson, N. N. Li, and D. E. O'Brien, U.S. Patent 4,397,354.
21. W. M. Salathiel, Muecke, C. E. Cooke, and N. N. Li, U.S. Patent 4,359,391 (Nov. 16, 1982).
22. C. R. Dawson, N. N. Li, and D. E. O'Brien, U.S. Patent 4,568,392 (Feb. 4, 1986).

23. R. P. Cahn, U. S. Patent 3,244,763 (April 5, 1966).
24. N. N. Li, *A.I.Ch.E.J.* *17*(2), 459–463 (1971).
25. N. N. Li, *Ind. Eng. Chem. Process Des. Dev.* *10*(2), 215–221 (1971).
26. T. A. Hatton, E. N. Lightfoot, R. P. Cahn, and N. N. Li, *Ind. Eng. Chem. Fundam.* *22*, 27–35 (1983).
27. W. E. Hardwick, U.S. Patent 4,221,658 (1980).
28. J. B. Scuffham, *Chem. Eng.* *370*, 328 (1981).
29. R. P. Cahn and N. N. Li, U.S. Patent 4,086,163 (April 25, 1978).
30. Alamine is an oil-soluble ion exchange resin marketed by Henkel Corp.
31. R. P. Cahn, U.S. Patent 4,839,056 (June 13, 1989).

7

Evaluation of Mass Transfer Coefficients from Single-Drop Models in Pulsed Sieve-Plate Extraction Columns

Qian Yu, Weiyang Fei, Lei Xia, and Jiading Wang *Tsinghua University, Beijing, China*

The pulsed sieve-plate extraction column (PSE column) is a column fitted with a stationary cartridge of horizontal sieve plates. The two phases pass though the column countercurrently with pulsation produced by a bellow or other pulsing devices. It has a great advantage over other mechanical agitated extractors while processing corrosive or radioactive solutions because there is no internal moving parts in PSE column. Therefore, it has received considerable attention as liquid–liquid extraction equipment for nuclear, hydrometallurgical, and chemical industry.

Extensive research work on the characteristics and the design of PSE column have been done in the past few decades. The operating characteristics of PSE column was described by Sege and Woodfield [1] and Geier [2]. There are five arbitrary operating regions, namely, mixer-settler region, emulsion region, instable operating region, flooding at insufficient and too high pulse intensity, which depend on the flow rate, and the degree of pulsation. It is understood that the emulsion region is most important for practical applications. Thornton [3], Logsdail et al. [4], Wang et al. [5], Zhu et al. [6], and many other workers have studied the hydrodynamics of the column. Some equations correlating the holdup and flooding velocities data have been proposed. The majority of data, however, were obtained in the absence of mass transfer. Many authors [7–10] have developed equations to correlate drop size. But the calculated values of d_{32} from different equations for given systems usually differ from each other. Besides, none of their measurements was carried out under the mass transfer condition.

In the early years, the mass transfer performance has been reported in terms of apparent volumetric mass transfer coefficient or apparent height of transfer unit (HTU_{app}) without correcting the effects of axial mixing [11,12]. Since the 1960s, massive research works have been carried out on the axial mixing in PSE column. Smoot [13] and Ziolkowski [14] published some "true" overall height of transfer unit (= $V_D/K_{OD}a$) results in which the effects of axial mixing have been considered and deducted. It was widely recognized in early 1980s that the evaluation of mass transfer and axial mixing parameters from measured concentration profiles in extraction columns might provide more reliable results [15–17]. Evaluation of "true" volumetric mass transfer coefficient and backmixing parameters of PSE column from the measured solute concentration profiles or by the dynamic stimulus-response technique were reported by Lei and coworkers [18]. Because the true mass transfer coefficient and interphase area are influenced by the physical properties of the systems and the operating conditions of the column in different ways, it is desirable to use separated correlations of these parameters for the design purpose. However, it is still difficult to find true mass transfer coefficient data in literatures.

In view of the above problems, this paper gives a fresh look and detailed study on the mass transfer of the PSE column. A systematic experimental study on mass transfer performance was carried out in a 40- and a 150-mm diameter PSE column operated with three typical systems. The concentration profiles of both phases, drop size distribution, and dispersed phase holdup were measured simultaneously. An improved computer program for estimating mass transfer and backmixing parameters from the concentration profiles was developed based on the backflow model. True mass transfer coefficients were obtained over a wide range of physical properties and operating conditions. Finally, a procedure based on new developed correlations of the mass transfer coefficient, the mean drop size, and the holdup has been worked out to put a step forward to the design of an extractor from the basic principles.

EXPERIMENTAL

The extraction systems used in present investigation are listed in Table 1. The aqueous phase was continuous (C) and organic phase was dispersed (D) in all runs [20].

Two glass columns were fitted with standard sieve-plate cartridge made of stainless steel. The principal dimensions of the columns were:

Column internal diameter	0.04 and 0.15 m
Column effective height	1.12 and 2.00 m
Compartment height	0.05 m
Diameter of holes	0.003 m
Fractional open area of plate	23%

Table 1 Systems Used in Experiments

System	Set	Direction of transfer	Interfacial tension (mN/m)	Diameter of column (mm)	Number of experiments	Symbol in figure	Diffusion coefficient	
							D_d 10^{-10}	D_c m²/sec
$H_2O/HNO_3/30\%$ TBP[a]	1	C→D	10.8	40	23	○	56.0	26.0
	2	C→D	10.8	40[b]	10	▽		
	3	C→D	9.8	150	17	■		
$H_2O/$succinic acid/ n-butanol	4	D→C	1.6	40	13	△	2.8	6.8
	5	C→D	1.2	40	13	▲		
$H_2O/HAc/30\%$ TBP	6	C→D	10.7	40	12	□	—	10.4

[a]TBP, Tri-n-butyl phosphate, kerosene as diluent.
[b]A pulsed column equipped with the dispersion–coalescence type cartridge [19].

Flow diagram of PSE columns is illustrated in Figure 1. Quasi-sinusoidal wave air pulsations were produced by compressed air through a three-way pulse valve controlled by a solenoid valve. Amplitudes were varied in the range of 5–15 mm and the frequency in the range of 1.0–2.5 sec^{-1}. Eight pairs of carefully designed samplers were located along the column to obtain the single-phase samples.

The photographs of drops were taken after the column had been operated about 50 min in order to ascertain the establishment of steady state operation. A camera equipped with extension tubes was used. The exposure time was 1/1000 sec when the ASA 400 B&W film was used. Samples of each phase were simultaneously taken from each sampling point 20 min later. Small capture caps packed with polypropylene fiber were used to obtain pure dispersed phase samples. After sampling, all the inlet and outlet valves of the column were rapidly closed and the holdup of dispersed phase was estimated by the displacement method.

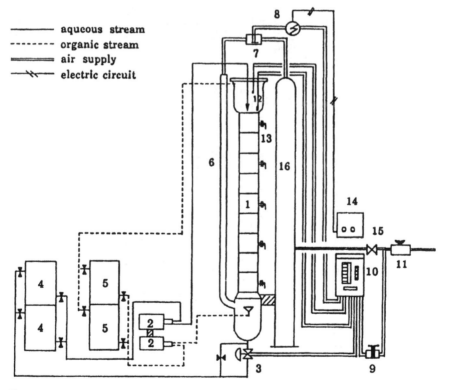

Figure 1 Flow diagram of 150 pulsed sieve-plate extraction column. (1) PSE column; (2) metering pumps; (3) control valve; (4) tank for water phase; (5) tank for organic phase; (6) pulsion leg; (7) three-way pulsion valve; (8) pulsion solenoid valve; (9) pressure regulator; (10) pneumatic cabinet; (11) pressure regulator; (12) pneumatic indicator; (13) sampling valve; (14) frequency regulator; (15) amplitude regulator; and (16) mast buffer tank.

A digitizer system was used to analyze the enlarged photographs of drops. Coordinates of drops on the pictures were sent from the digital pad to a microcomputer. A statistics program was developed to get the drop size distribution curves and the corresponding statistic mean drop size, which was based on over 350 drop readings for one mean size. An optical correcting function was deduced to compensate the optical distortion on drop photographs caused by the curvature of the column wall and the refraction in different mediums. Seventy sets of drop mean size and distribution data were obtained in total.

Acid concentration of samples were analyzed with an autopipetting digital titration system. The data showed a good reproducibility. Accumulative errors in measurements were about 2% on average. Eighty-eight sets of concentration profiles were obtained for three systems in two mass transfer directions as illustrated in Table 1.

DROP SIZE AND DISTRIBUTION

Mean drop size in an extraction column is an essential design parameter which affects both the hydrodynamic and mass transfer performance of the column. Therefore, it should be considered as the starting point in the design of a solvent extraction column from basic principles.

For a measured drop size distribution, the volume fraction function can be calculated from:

$$f(d_i)\Delta d_i = \frac{n_i d_i^3}{\sum_j n_j d_j^3} \tag{1}$$

The expectation value and standard deviation of the distribution function $j(d)$ are deduced as follows:

$$M(d) = \sum d_j f(d_j)\Delta d_j = \frac{\sum n_j d_j^4}{\sum n_j d_j^3} = d_{43} \tag{2}$$

$$D(d) = \sum (d_j - M(d))^2 j(d_j)\Delta d_j = d_{53}^2 - d_{43}^2 \tag{3}$$

Various distribution equations were examined to describe the drop size distribution in PSE columns. The arithmetic normal distribution equation was found to be suitable for the drop size distribution in PSE columns. It has a good reliability in the parameter estimation too. From the moment analysis, the equation and parameters can be derived as:

$$f(d) = \frac{1}{\sqrt{2\pi}\sigma} \exp\left\{-\frac{(d - d_{43})^2}{2\sigma^2}\right\} \tag{4}$$

$$\sigma^2 = d_{53}^2 - d_{43}^2 \tag{5}$$

Mean drop size, d_{43}, was determined from the balance between the surface energy

of drops and the turbulent energy which drops obtained in the extractor. And it
turned out, the latter has much to do with the consumption mechanisms of input
pulsation energy [14]. From these the following equation could be derived:

$$d_{43} \propto \left(\frac{\gamma}{\rho_c}\right)^{\frac{3}{5}-\frac{8}{5}} (A_f)^{-\frac{1}{5}\sim-\frac{8}{5}} \tag{6}$$

Summarizing our 60 data for two systems ($\gamma = 1.2$–10.8 mN/m) and Ugarcic's
seven data [21] for the oxylene/acetone/water system ($\gamma = 32$ mN/m), the
correlation of mean drop size was obtained as the following [22]:

$$d_{43} = 0.0123 \left(\frac{\gamma}{\rho_c}\right)^{0.43} (Af)^{-0.70} \tag{7}$$

The standard relative error of this equation is about $\pm 14\%$. Equation 5 can be
transformed into a more convenient linear form and be correlated as

$$\sigma = 0.31 d_{43} \tag{8}$$

DISPERSED PHASE HOLDUP AND INTERFACIAL AREA

Some empirical correlations for predicting of holdup in PSE columns have been
proposed [23]. However, the empirical constants in these correlations are not
always available for a new system or column geometry. In this chapter a new
equation is derived from the single-drop-movement mechanism to correlate the
drop size, holdup, and throughput. Different flow regimes in PSE columns have
been observed. In the emulsion regime, Reynolds numbers are in the range of 2–
500. The Allen velocity equation (Eq. 24, in Table 4) could be used to describe a
single drop moving through an infinite fluid. As to liquid drop, the inner
circulation should be considered and the velocity equation modified with a factor
of $(3\mu_d + 3\mu_c)/(3\mu_d + 2\mu_c)$ [24]. There is also an interaction between drops [25]
and the resistance of the internals to drop movement in extraction columns, so that
the slip velocity $u_s(d)$ is expressed as:

$$u_s(d) = C_R \left(\frac{3\mu_d + 3\mu_c}{3\mu_d + 2\mu_c}\right) \left(\frac{4\Delta\rho^2 g^2}{225\rho_c u_c}\right)^{\frac{1}{3}} d(1 - \Phi)^n \tag{9}$$

The overall material balance of the dispersed phase should be satisfied:

$$\sum_j u_s(d_j)f(d_j)\Delta d_j = \frac{V_D}{\Phi} + \frac{V_C}{1 - \Phi} \tag{10}$$

Inserting Eq. 9 into Eq. 10, a hydrodynamic equation of swarm of drops in
extraction columns could be obtained as follows:

$$\frac{V_D}{\Phi} + \frac{V_D}{1 - \Phi} = C_R \left(\frac{3\mu_d + 3\mu_c}{3\mu_d + 2\mu_c}\right) \left(\frac{4\Delta\rho^2 g^2}{225\rho_c u_c}\right)^{\frac{1}{3}} d_{43}(1 - \Phi)^n \tag{11}$$

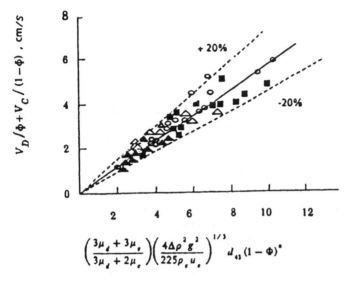

$$\left(\frac{3\mu_d + 3\mu_e}{3\mu_d + 2\mu_e}\right)\left(\frac{4\Delta\rho^2 g^2}{225\rho_c u_c}\right)^{1/3} d_{43}(1-\Phi)^n$$

Figure 2 Correlation of the hydrodynamic equation of swarm of drops in PSE columns (for symbols, see Table 1).

The constriction factor C_R is considered to be a variable for different column geometries. As to PSE columns with standard cartridge, Eq. 11 was plotted in Figure 2. It was found out that the $C_R = 0.57$ is the best fit of the equation for the experimental data. The exponent n could be estimated from a correlation developed by Zhu et al. [6].

From Eq. 11, dispersed phase holdup Φ can be calculated. Then the interfacial specific area is calculated as:

$$a = \sum_j n_j(\pi d_j^2)\frac{\Delta d_j}{\sum_j n_j}\left(\frac{\pi}{6}d_j^3\right)\frac{\Delta dj}{\Phi} = 6\Phi\frac{\sum_j n_j d_j^2}{\sum_j n_j d_j^3} = \frac{6\Phi}{d_{32}} \tag{12}$$

A link between the volume/surface mean size d_{32} and the volume fraction mean size d_{43} is very useful. Comparing the drop data, it is obtained as follows:

$$d_{32} = 0.91d_{43} \tag{13}$$

BACKFLOW MODEL AND PARAMETER ESTIMATION

The backflow model is a wide used model which takes into account the axial mixing of two phases in an extraction column. It is close in concept to the pulsed sieve-plate column and convenient for computation. The model is described by the following set of equations:

$$\alpha_c x_{j-1} - (1 + 2\alpha_c)x_j + (1 + \alpha_c)x_{j+1} - \frac{R_j V_D}{V_C} = 0 \qquad (14)$$

$$(1 + \alpha_D)y_{j-1} - (1 + 2\alpha_D)y_j + \alpha_D y_{j+1} + R_j = 0 \qquad (15)$$

$$R_j = \frac{K_{OD,j}a(y_j^* - y_j)H}{NV_D} \qquad j = 2, 3, \cdots, N - 1$$

For the 1st and Nth stage, the closed boundary condition is assumed. Given V_c, V_D, a, and $y_0 = y_{in}$, $x_{N+1} = x_{in}$, the equations include 2N variables (x_j, y_j, j = 1, . . ., N) and three parameters (the interphase mass transfer coefficient K_{OD}, the backflow ratios in both phases α_C and α_D). An improved computation algorithm was proposed to calculate the concentration profiles x^{cal}, y^{cal} for the systems with nonlinear phase equilibrium behavior.

A maximum likelihood objective function was constructed from comparing the concentration profiles x^{cal} and y^{cal} with the measured ones x^{mea} and y^{mea}. The model parameters (K_{OD}, α_C, α_D) can be evaluated by the procedure of optimum parameter estimation for the multivariable nonlinear equations. The comparison between the calculated concentration profiles and the measured ones is shown in Table 2. The random error analysis indicates that the deviations of the estimated parameters are 3% for K_{OD} and 10% for α_c, α_D when a deviation of 2% in concentration profile measurement is imposed.

Table 2 Comparison of Measured and Calculated Concentration Profiles

N	x^{med}	x^{cal}	y^{med}	y^{cal}
x_{in}	2.156			
1	2.113	2.118	0.459	0.465
6	2.082	2.080	0.449	0.454
16	1.985	1.982	0.426	0.426
26	1.853	1.848	0.348	0.388
37	1.659	1.641	0.335	0.329
47	1.341	1.381	0.259	0.255
57	1.040	1.056	0.141	0.145
60	0.974	0.994	—	—
y_{in}			0.068	

Effective height of column, 2.0 m; column diameter, 150 mm; Af = 10.6 mm/sec; System: water/nitric acid/30% TBP; V_c = 1.2 mm/sec; V_D = 3.35 mm/sec; Standard deviation in concentration profiles simulation: 0.014 mol/liter.

"TRUE" INTERPHASE MASS TRANSFER COEFFICIENT IN PSE COLUMNS

Seventy sets of interphase mass transfer coefficients for two experimental systems were obtained. They were in both mass transfer directions and at different levels of pulsation intensities, with the standard sieve plate cartridge and the "dispersion-coalescence" cartridge, in 40- and 150-mm-diameter columns. Some typical results are listed in Table 3. It seems that the "true" interphase mass transfer coefficient can be predicted by the well-known single-drop mass transfer models (Table 4). In emulsion operation regime, drop Reynolds number is mainly in the range of 10–200, where the turbulent circulating pattern dominates inside the drop. Handlos and Baron [28] considered the case of a turbulent drop with the circulation pattern simplifies to concentric circles. Liquid between two streamlines in drop becomes fully mixed after one circuit. K_d can be calculated by

$$Sh_D = 0.00375 Pe_D = \frac{0.00375 V_s d_{32}}{D_d(1 + \mu_d/\mu_e)} \tag{18}$$

It should be pointed out that the main resistance to mass transfer is located in drop side in this investigation. Therefore, the choice of mass transfer model in continuous phase is less sensitive to the magnitude of the overall coefficient. The Bossinesd equation (Eq. 18) could be used to describe the mass transfer in continuous phase [28,35].

The interphase mass transfer coefficient K_{OD} is calculated from individual film coefficients k_c and k_d on the concept of additivity of resistances. A reasonable agreement was found in comparison between the K_{OD} from the model and from the experimentals over a wide range of Pe_D from 800 to 80,000. The average deviation is about 17%. The results are illustrated in Figure 3 in terms of dimensionless groups Sh_D, namely Pe_D.

The mass transfer coefficients were found to be directly proportional to the slip velocity V_s, or mean drop size d_{43}. It follows the same trend as predicted by the Handlos–Baron model and has been verified by many workers in other types of liquid extractors [20]. The evidence is obviously against the application of either the rigid drop model or the Kronig–Brink mass transfer model because these two models predict a decrease in mass transfer coefficient with the increase of drop size.

The system of water/succinic acid/n-butanol involved the transfer of succinic acid between butanol drops and continuous water solution in both directions. It was noticed that the mass transfer Sh_{OD} numbers for the C→D direction were about 30% higher than those for the D→C direction. This dependency of the mass transfer coefficient on the direction of solute transfer may be considered the effect of the interfacial tension gradient in adjacent region of drops [36].

Table 3 "True" Mass Transfer Coefficients in Pulsed Sieve-Plate Extraction Columns and Comparison of Measured with Calculated Sh_D Numbers

Set	No.	A.f (mm/sec)	V_c (mm/sec)	V_D (mm/sec)	Φ	d_{43} (mm)	K_{OD} (10^5 m/sec)	$Pe_D \times 10^3$	Sh_{OD}^{exp}	Sh_{OD}^{cal}
1	13	10 1.1	1.36	3.54	0.08	2.36	4.35	6.36	19.4	20.8
	14	1.1	1.18	4.12	0.08	2.10	4.62	6.25	17.4	21.2
	15	1.3	1.36	3.54	0.08	2.22	4.69	5.51	19.6	18.1
	16	1.5	1.19	3.34	0.08	1.55	3.59	3.75	9.9	13.0
	17	2.0	1.19	3.34	0.07	1.37	4.30	3.74	10.5	13.0
	18	2.5	1.59	4.60	0.17	1.19	2.86	1.96	6.27	6.83
	19	2.5	1.21	3.40	0.12	1.06	2.78	1.85	5.56	6.32
2	51	10 2.5	1.20	3.40	0.186	1.05	2.79	1.16	5.2	4.2
	52	2.0			0.122	1.18	3.15	1.92	6.6	6.8
	53	1.7			0.107	1.49	3.19	2.74	8.5	9.6
	54	1.5			0.099	2.08	4.36	4.13	16.2	14.3
	55	1.3			0.101	2.86	4.78	5.58	24.4	19.0
	56	1.1			0.092	3.17	5.52	6.76	31.3	22.8
3	74	10 2.0	1.20	3.35	0.180	1.05	3.04	1.49	5.7	5.4
	75	1.8			0.160	1.05	2.55	1.67	4.8	6.1
	76	1.6			0.123	1.35	3.73	2.74	9.0	9.9
	73	1.5			0.132	1.33	4.06	2.53	9.7	9.2
	72	1.3			0.090	1.80	6.40	4.93	20.6	17.6
	77	1.3			0.090	1.72	4.71	4.71	14.5	16.9
	78	1.2			0.080	2.14	6.59	6.55	25.1	23.3
	79	1.1			0.071	2.41	6.77	8.30	29.2	29.3
4	106	5 1.00	2.00	2.00	0.065	1.80	2.74	60.4	175	177
	108	1.17			0.072	1.61	2.51	49.3	144	148
	110	1.33			0.082	1.33	2.07	36.1	98	111
	112	1.50			0.105	1.13	1.70	24.5	68	77
5	107	5 1.00	2.00	2.00	0.098	1.35	1.88	31.3	90.3	97.4
	109	1.17			0.101	1.17	2.28	26.4	95	83
	111	1.33			0.119	1.11	2.17	21.7	86	70
	113	1.50			0.143	0.99	1.88	16.5	66	54

DESIGN OF PSE COLUMN FROM FIRST PRINCIPLES

Summarizing the result obtained by the author and others, a new design procedure of pulsed sieve-plate columns is proposed from the basic principles of extraction process. A sketch map of the design method is illustrated in Figure 4. Provided with physical properties, a phase equilibrium equation of the system, and operating conditions of the column, drop size distribution and mean size can be

Table 4 Single-Drop Mass Transfer and Movement Models

Drop state	Re	Mass transfer in drop	Mass transfer in continuous phase	Movement of drop [33]
Stagnant	0–1	Newman [26]: $Sh_d = \frac{2}{3}\pi^2$ (16)	Steinberger and Treybal [30]: $Sh_C = 2.0 + 0.569(Gr\cdot Sc)^{\frac{1}{4}} + 0.347(Re\cdot Sc^{\frac{1}{2}})^{0.62}$ (20)	Stokes law (Re < 1) $u_i(d) = \frac{d^2\Delta\rho g}{18\mu_c}$ (23)
Laminar circulating	2–10	Kronig and Brink [27]: $Sh_d = 17.9$ (17)	Bonssinesq [31]: $Sh_c = \left(\frac{4}{\pi}Re\cdot Sc\right)^{\frac{1}{2}}$ (21)	Allen law (Re = 2–500) $u_i(d) = \left(\frac{4\Delta\rho^2 g^2}{225\rho_c\mu_c}\right)^{\frac{1}{3}}d$ (24)
Turbulent circulating	10–200	Handlos and Baron [28]: $Sh_d = 0.0037Pe_d$ (18)		
Oscillating	>200	Rose and Kintner [28]: $k_d = 0.45(wD_d)^{\frac{1}{2}}$ (19)	Garner and Tayeban [32]: $Sh_c = 50 + 0.0085(Re\cdot Sc^{-0.7})$ (22)	Newton law (Re > 500) $u_i(d) = (\Delta\rho g/\rho_c)^{\frac{1}{2}}$ (25)

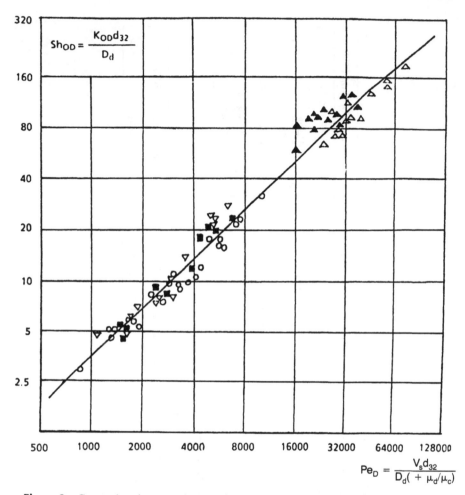

Figure 3   Comparison between the experimental and the predicted values of the interphase mass transfer coefficients. ————, predicted from the single drop mass transfer models. For symbols, see Table 1.

calculated by Eqs. 4 and 7, which is derived from the energy balance mechanism of drops. Flooding point, dispersed phase holdup, and interfacial area can be calculated by the hydrodynamic equation of drops (Eqs. 11 and 12). Interphase mass transfer coefficient is calculated from the turbulent circulating models (Eqs. 18 and 21). Finally, axial mixing parameters are correlated with column diameter and hydrodynamic condition. The height of extraction column to fulfill a given separation demand can be calculated from the backflow model. An example of column height calculation and the comparison with the actual height of the column

Optimization of operating condition

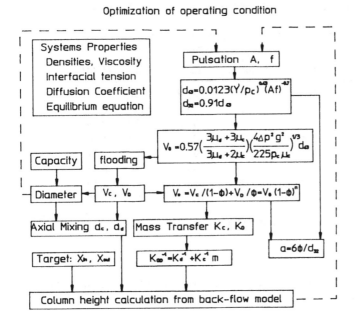

Figure 4   Sketch map of design method of pulsed sieve-plate extractor.

are shown in Table 5. The calculated column height was found to be close to the experimental one. Furthermore the optimization of operating conditions can be carried out to seek a higher efficiency for a given situation.

It is found out that the K_{OD} is more or less independent of the kinds of cartridge and column diameter, which shows that the K_{OD} is a parameter reflecting the transport mechanism inside the drop and adjacent boundary regions. There are no obvious differences between the 40- and 150-mm diameter columns for the data of drop size and interfacial specific area too. These results are very important for scaling up a pulsed sieve-plate extraction column directly from small-scale experiments in the case of lacking pilot plant data.

CONCLUSIONS

In the emulsion operating regime of pulsed sieve-plate columns with standard cartridge, drop size distribution can be described by the normal probability function. Mean drop size is correlated as

$$D_{43} = 0.0123 \left(\frac{\gamma}{\rho_e}\right)^{0.43} (Af)^{-0.70}$$

Table 5 Comparison of Calculated Height of Extraction Column with Actual Height

No.	A (mm)	f (liter/sec)	V_c (mm/sec)	V_D (mm/sec)	$K_{OD}a$, 10^{-3}/sec Exp.	Cal.	H^{cal}/H
127	10	1.00	1.65	3.33	5.39	5.66	1.08
128		1.25			5.69	6.90	0.87
129		1.50			8.03	8.20	0.79
134		1.00	3.00	6.00	9.60	9.32	1.20
135		1.63			18.0	14.8	1.11
136		1.50	2.35	4.70	10.6	11.7	0.97
137		1.50	3.00	6.00	15.4	15.7	1.10
138		150	3.65	7.30	16.3	17.5	1.18

Note: System, H_2O/Hac/30% TBP. Effective height of the column, 1.12 m.

A hydrodynamic equation of swarm of drops is given as the following, from which the dispersed phase holdup and interfacial specific area can be calculated.

$$\frac{V_c}{1 - \Phi} + \frac{V_D}{\Phi} = 0.57 \left(\frac{3\mu_d + 3\mu_c}{3\mu_d + 2\mu_c}\right)\left(\frac{4\Delta\rho^2 g^2}{225\rho_c u_c}\right)^{\frac{1}{4}} d_{43}(1 - \Phi)^n$$

The "true" interphase mass transfer coefficients K_{od} in the pulsed sieve-plate extraction column were experimentally obtained from the concentration profiles of two phases: the dispersed phase holdup and the drop size. It is found from this work that the K_{od} could be predicted reasonably from the single-drop turbulent circulating models.

These results have been used for developing a computer program to put a step forward to the design of an extractor from the basic principles.

ACKNOWLEDGMENT

We acknowledge the support of the National Natural Science Foundation of China on this work.

NOMENCLATURE

A   pulse amplitude (peak to peak), m
a   interfacial specific area, m^2/m^3
C_R constriction factor
D   diffusivity coefficient, m^2/sec
d   drop diameter, m

d_{32} volume/surface (Sauter) mean size, m
d_{43} drop volume fraction mean size, m
f pulse frequency, m^{-1}
$f(d)$ drop volume fraction distribution function, m^{-1}
H effective height of column, m
K_{OD} "true" interphase mass transfer coefficient based on dispersed phase, m/sec
k mass transfer coefficient in individual phase, m/sec
N number of stages in backflow model
Pe_D mass transfer Peclet number $V_s d_{32}/D_d(1 + \mu_d/\mu_c)$
Re drop Reynolds number $d_{32}V_s\rho_c/\mu_c$
Sc Schmidt number $\mu c/\rho_c D_c$
Sh mass transfer Sherwood number kd_{32}/D
u_t drop terminal velocity, m/sec
V superficial velocity, m/sec
x, y concentration of solute in continuous and dispersed phases, mol/liter
x backflow ratio
Φ dispersed phase holdup
ρ density, kg/m
μ viscosity, kg/m·sec
γ interfacial tension, N/m
σ standard deviation of distribution, m

Affixes
C   continuous phase
D   dispersed phase
d   drop
s   slip (relative to continuous phase)

REFERENCES

1. G. Sege and F. W. Woodfield, *Chem. Eng. Prog. 50*(8), 396 (1954).
2. R. G. Geier, *USAEC Report* TID-7534, Book 1, 107 (1957).
3. J. D. Thornton, *Brit. Chem. Eng. 3*, 247 (1958).
4. D. H. Logsdail and J. D. Thornton, *Trans. Inst. Chem. Eng. 35*, 331 (1957).
5. Wang Jiading, Sheng Zhongyou, and Wang Chemfan, *J. Chem. Ind. Eng. (Chinese) 16*, 216 (1965).
6. S. Zhu, B. Zhang, Z. Shen, and J. Wang, *J. Chem. Ind. Eng. (Chinese) 33*(1), 1 (1982).
7. S. Z. Kagan, et al., *Int. Chem. Eng. 5*, 656 (1956).
8. T. Misek, *Coll. Czech. Chem. Commun. 28*, 570, 1963.
9. S. Z. Kubica, et al., *Inz. Chem. (Poland) 7*(4), 903 (1977).
10. T. Miyauchi and H. Oya, *AIChEJ. 11*, 395 (1965).
11. J. D. Thornton, *Trans. Inst. Chem. Eng. 35*, 316 (1957).

12. L. D. Smoot, B. W. Marr, and A. L. Babb, *I&EC 51*, 1005 (1959).
13. L. D. Smoot and A. L. Babb, *I&EC Fundam. 1*, 93 (1962).
14. Z. Ziolkowski, et al., *Inz. Chem. (Poland) 4*(1), 163 (1974).
15. F. L. Spencer, et al., *AIChEJ. 27*, 1008 (1981).
16. N. L. Ricker, F. Nakashio, and C. J. King, *AIChEJ. 27*, 277 (1981).
17. V. Rod, W. Y. Fei, and C. Hanson, *Chem. Eng. Res. 61*, 290 (1983).
18. X. Lei, Y. Qian, W. Y. Fei, and J. D. Wang, *Proc. ISEC 3*, 212 (1986).
19. Y. Qian, W. Y. Fei, and J. D. Wang, *J. Chem. Ind. Eng. (Chinese) 39*(5), 522 (1988).
20. Y. Qian, Interphase Mass Transfer Process in Pulsed Sieve Plate Extraction Columns, Ph.D. dissertation, Tsinghua University, Beijing, 1987.
21. M. Ugarcic, Ph.D. dissertation, TCL, Swiss Federal Inst. of Tech., 1981.
22. Y. Qian, W. Y. Fei, and J. D. Wang, *J. Tsinghua Univ. (Chinese) 28*(6), 16 (1988).
23. A. Kumar and S. Hartland, *Chem. Eng. Res. Dev. 61*, 248 (1983).
24. W. Rybezinski, *Bull. Acad. Cracovie*, Ser. A, 40 (1911).
25. Y. Y. Dai, X. Lei, and J. D. Wang, *J. Chem. Ind. Eng. (Chinese) 39*(4), 430 (1988).
26. A. B. Newman, *Trans. AIChEJ. 27*, 310 (1931).
27. R. Kronig and J. C. Brink, *Appl. Sci. Res. A2*, 143 (1950).
28. A. E. Handlos and T. Baron, *AIChEJ. 3*, 127 (1957).
29. P. M. Rose and R. C. Kintner, *AIChEJ. 12*, 530 (1966).
30. R. L. Steinberger and R. E. Treybal, *AIChEJ. 6*, 226 (1960).
31. J. Boussinesq, *J. Math. Pure Appl. 11*, 285 (1905).
32. F. H. Garner and M. Tayeban, *Anal. Real. Soc. Espan. Fis. Quim. (Madrid) B56*, 479 (1960).
33. D. Kunni and D. Levenspiel, *Fluidization Engineering*, Wiley, New York, 1969.
34. Y. F. Su, et al., *Proc. Inter-Collegiate Symp. Chem. Eng. (Chinese) 47* (1963).
35. F. J. Zuiderweg and A. Harmens, *Chem. Eng. Sci. 9*, 89 (1958).

8

Liquid Waste Concentration by Electrodialysis

Rémy Audinos   *Ecole Nationale Supérieure de Chimie, Institut National Polytechnique, Toulouse, France*

INTRODUCTION

Liquid effluents are one of the main sources of pollution attributable to many industries: inorganic and organic chemicals, foods, hydrometallurgy, nuclear, petroleum, petrochemical, pharmaceutical, surface treatment, etc.

Normally, these effluents are produced continuously. Besides this type of pollution, which is closely linked to production, there exist other intermittent sources of pollution due to purging. Purging is needed to maintain the concentration of harmful ions below a certain critical value essential to keep a reaction (biochemical, chemical, electrochemical, or nuclear) at an appropriate level to achieve economic yields.

These effluents are sources of pollution. This pollution is direct and evident if wastes are discarded directly into the environment. Also sometimes, in order to reduce pollution, one obtains the opposite. One example is the neutralization of acid effluents with lime milk. Instead of chemical pollution, there is mechanical pollution due to the quarry where the limestone is extracted and to the dump where the calcium salt is stored, not to mention the transportation between the three sites: quarry, plant, and dump.

On the other hand, apart from being a source of pollution, these effluents are the cause of the loss of a considerable amount of raw biochemical, chemical, and metallic material. For example, in France, losses from surface treatments alone for 1980, all metals taken into account, were in the range of 1000 metric tons per year, i.e., over 0.5% of the total tonnage used as indicated in Table 1.

Table 1 Amounts of Metals Lost in Various Wastes in France in 1980

Metal	Use	Amount used (tons/year)	Amount lost (tons/year)	Recovery
Cadmium	Electrodeposition	250	25	None
Copper	Electrodeposition	2,000	140	Small
	Pickling	160,000	200	Great
Nickel	Electrodeposition	3,000	150	Small
Silver	Jewelry	40	—	Great
	Industry	30	—	Great
	Photography	200	15	Great
Tin	Electroplating	4,500	30	Small
	Electrodeposition	200	20	Small
Zinc	Electrodeposition	3,000	250	Very small

Source: Data from Ref. 1.

In the chemical industries these wastes issue from the plant where organic or inorganic compounds are produced from raw materials or from fine chemicals. For example, in some organic or inorganic chemical processes there is a neutralization step after the reaction, such that large amounts of neutral salts are produced in dilute solution.

In many cases, sodium sulfate is the principal pollutant in aqueous effluents from chemical processes, such as rayon production, organic acid and silica manufacture, organic synthesis, or flue gas desulfurization. As another example, in 1980 the worldwide production of rayon was ~3×10^9 tons, and approximately 500 m^3 of water per kg of rayon is needed directly for processing, and 8–10 times that amount of water is required to provide supplementary service facilities for the plant [2]. The resultant effluent solutions are more or less dilute. To avoid discharging sodium sulfate wastes into lakes, rivers, or the sea, it is often necessary to carry out a preconcentration step via electrodialysis, before complete evaporation or other physical or chemical treatment, such as acid and base regeneration (e.g., see section "Membrane Electrohydrolysis").

In hydrometallurgy plants, liquid wastes are obtained when ores are treated by liquor to extract the metal. For example, the production of 1 metric ton of titanium dioxide gives rise to approximately 3 metric tons of sulfuric acid (100% H_2SO_4), and 2 metric tons of ferrous sulfate ($FeSO_4$, $7H_2O$) in a waste solution containing an average of 20% of free sulfuric acid and 10–12% of transition metals and alkaline earth metals (cf. Table 2). Since worldwide production is a little less than 3 million tons of titanium dioxide per year, this requires about 10 million tons of sulfuric acid. But this acid is wasted, as the cost of H_2SO_4 regenerated from the effluent is higher than the cost of freshly prepared H_2SO_4.

Table 2 Wastes Generated by TiO_2 Production in Some Member Countries of the Economic Commission for Europe

Country	Quantity of material for disposal			
	Acid solution		Ferrous sulfate	
	As 100% H_2SO_4 (10^3 ton/year)	Ton per ton of TiO_2 produced	As $FeSO_4$, $7H_2O$ (10^3 ton/year)	Ton per ton of TiO_2 produced
Belgium	140	2.00	105	1.50
Czechoslovakia	25	1.00	129	5.16
Finland	124	1.55	360	4.50
France	370	2.07	237	1.33
Germany, Fed. Rep.	371	1.46	402	1.58
Italy	132	2.24	33	0.65
Netherlands	70	2.01	15	0.83
Norway	37	1.49	43	1.71
Poland	72	2.03	120	3.40
Spain	100	1.90	197	3.77
United Kingdom	220	1.73	165	1.30

Source: Data from Ref. 3.

As a consequence, any improvement in the extraction efficiency of metal will reduce the mineral content in waste effluents. This mineral content is due to the presence of mineral ions.

After production of raw metal there is always a treatment step in the plant or elsewhere to yield material suitable for the metallurgical industries. In surface treatment processes, effluents are discharged either when metals are produced or when the surface of metals plastics, glass, etc., are treated for some special reason. Generally, these surface treatments come under three major headings: pretreatment, surface modification, and metal deposition, as indicated in Tables 3–6. Pretreatment involves surface preparation of metal, plastic, glass, etc., for the step which follows as indicated in Table 3. Surface modification generally involves a thin deposit on the surface to produce improved operability in the next step, or to protect previously deposited material, as indicated in Table 4.

In some cases, the change of the superficial aspect is the last step before use, as indicated in Table 5. But the primary purpose of surface treatments is the deposition of a metal on the surface of other metals or plastics, as indicated in Table 6.

In nuclear plants, radioactive wastes are the subject of special treatment, but some nonradioactive effluents can also be discharged into the environment. These

Table 3 Pretreatment Operations in Surface Treatment

Operation	Purpose	Substrate	Bath
Activation	Sensitization of the surface to be treated	Plastics	Acids or cyanides and metallic salts
Degreasing	Elimination of oil, grease	All kinds of metals, plastics	Bases, or bases and cyanide, or acid and additives (tensio-actives, ligands)
Etching	Elimination of surface oxides	All kinds of metals, plastics	Acid or bases and additives (tensio-actives)
Neutralization	Change of surface pH	All kinds of metals, plastics	Bases and cyanides or acids

Source: Data from Ref. 4.

effluents generally arise from chemical additives in the secondary circuit. For example, in the PWR (pressure water reactor), the boron content of the secondary circuit is maintained at a constant level by neutralization with a base. The nonradioactive salts formed are fixed on ion exchange resins. Then the eluted solutions, containing most of the salts, are discarded.

Membrane techniques, and especially electromembrane processes such as electrodialysis, electro-electrodialysis, or diffusion processes such as Donnan dialysis, or pressure-driven processes such as reverse osmosis, provide ways of concentrating and recovering these ions.

But it is a common mistake not to distinguish between membranology, even membranes themselves, and membrane techniques. However, the latter are bound in an inseparable trilogy, each absolutely necessary, since a membrane on its own is not capable of behaving as a selective barrier. The *method* arises from the intersection of three scientific fields related to membranes, to energies, and to mixtures, as previously defined [5]. These three domains of knowledge must be well understood in order for someone to be able to operate or to design a plant. Moreover, this basic membrane–energy–mixtures intersection (MEM) dictates the permeability and the selectivity of the basis element. Subsequently, the *process* embodies this intersection and it is a logical follow-on, from the basic element to the modules and finally to the system.

THE FUNCTIONS OF ELECTRODIALYSIS

Purpose of Electrodialysis

Electrodialysis (ED) is one of the techniques widely used for demineralization, e.g., desalination of water, demineralization of cheese wheys, and concentration of

Table 4 Principal Surface Modifications

Operation	Purpose	Substrate	Bath
Activation	Sensitization of the surface to be treated	Plastics and metallic salts	Acids or cyanides
Brightening	Preparation of a bare surface for the following treatment	Mainly aluminum or stainless steel	Acids or chromium and acids
Electropolishing	Improvement of surface properties	All kinds of metals	Acids or chromium and acids
Oxidation	Protection against corrosion before painting with a coat of metal oxide	Aluminum	Acids or chromium and acids
Passivation	Protection of the deposit	All kinds of metals	Acids or bases and chromium and additives (organic compounds)
Phosphating	Protection against corrosion before painting with a coat of metallic phosphate	Steel, aluminum, zinc	Zn, Fe, Mg phosphates and additives (tensio-actives, organic compounds)
Silting	Protection against corrosion by hydrolysis of the metallic oxide surface layer	Aluminum	Metallic salts and additives (dyes)

Source: Data from Ref. 4.

Table 5 Principal Superficial Changes

Operation	Purpose	Substrate	Bath
Chemical machining	Making a given form weight reduction	Metals	Acids or bases
Colorization	Decoration	Metals	Metallic salts, dyes
Demetalization	Retouching by metal relieving	Metals	Acids or bases and cyanides and organic compounds
Electroerosion	Making a given form weight reduction	Metals	Acids or bases
Surfacing	Making a nonreflecting surface	Metals, glass	Metallic salts, acids and salts

Source: Data from Ref. 4.

Table 6 Principal Metal Deposits

Operation	Purpose	Substrate	Bath
Noble metals (Ag, Au)	Anticorrosion, decoration, change of surface conductivity	Metals, plastics, ceramics	Metal and bases, metal and acids, metal and acids, and additives
Alloy deposit	Anticorrosion, decoration, change of surface conductivity	Metals, ceramics	Metals and acids, metals and bases and cyanide
Cadmium plating (Cd)	Preparation for anticorrosion, antifriction, weldability, ductility	Metals	Cadmium and cyanides, cadmium and acids and additives
Cyaniding (CN)	Increase in hardness	Steel	Salts and cyanides
Chromium plating (Cr)	Anticorrosion, decoration, hardening, antiabrasion, antifriction	Metals, plastics	Chromium and acids and additives
Coppering (Cu)	Preparation for electroforming	Copper alloys, zamak, plastics	Cu and acids; or Cu and bases and cyanides and additives (brighteners, tensioactives, organic compounds)
Nickel plating (Ni)	Preparation for anticorrosion protection, decoration, electroforming	Metals, plastics	Nickel and acids, or nickel and bases and additives
Tinning (Sn)	Anticorrosion, change of surface	Metals	Tin and bases, tin and acids, tin and additives
Galvanizing (Zn)	Preparation for anticorrosion protection	Steel	Zinc and acids, or zinc and bases, zinc and cyanides, zinc and additives

Source: Data from Ref. 4.

seawater prior to evaporation [6]. The use of ion exchange membranes to concentrate seawater and to produce freshwater from saline water has been investigated for about a third of a century and has now reached the practical stage. The entire domestic supply of table salt in Japan is produced by electrodialysis, and there are currently 11 plants of this type in Japan and other Asian countries (cf. Table 7). Many successful applications concerning the production of freshwater from the sea have been made in the United States, the Mediterranean, and Middle Eastern countries. Plants for the production of freshwater for use in boilers have been constructed in Japan, and a plant for the recovery of metals in concentrated brine from a reverse osmosis freshwater plant exists in Saudi Arabia.

Table 7 Salt-Manufacturing Plants Using Electrodialysis

Name of company	Capacity (millions of tons of salt per year)	Technology	Location
Shin Nihon Chem Ind Co.	224,000	Asahi Chemical	Japan
Ako Sea Water Chem Ind Co.	230,000	Asahi Chemical	Japan
Naruto Salt Mfg Co.	212,000	Asahi Chemical	Japan
Naikai Salt Works	192,000	Asahi Glass	Japan
Sakito Salt Mfg Co.	179,000	Asahi Glass	Japan
Kinkai Salt Mfg Co.	173,000	Tokuyama Soda	Japan
Sanuki Salt Mfg Co.	172,700	Tokuyama Soda	Japan
Tsung-Hsiao Electrodialysis Salt Factory	116,000	Asahi Chemical	Taiwan
Hanju Corporation	139,000	Asahi Chemical	Korea
Total	1,637,000		

Source: Data from Refs. 7 and 48.

Knowing that the membrane area of the Ako plant is of the order of 100,000 m^2, the total surface area of membranes used for the concentration of seawater is about 700,000 m^2.

But electrodialysis is also widely used in the field of waste treatment. It is now frequently used for the purification of effluents arising from the treatment of metallic surfaces or from chemical plants.

The basic concept of the MEM intersection, which was previously defined and characterized [5], is based on applying electrical energy to a medium containing ions in an aqueous solution, so that separation occurs in the presence of ion exchange membranes. Used to recover metals in the form of ions in solution, this process enables low-ion-concentration solutions to be concentrated [8–10]. Nevertheless, unlike pressure-driven processes, such as reverse osmosis, ED has the advantage of linking energy expenditure to the quantity of electrolytes to be extracted, and not to the volume of water to be treated. Unlike diffusion processes, such as dialysis, ED has the advantage of recovering ions in solution at the desired concentrations, and generally with the concentrating solution more concentrated than the feed solution. Unlike adsorption processes, such as ion exchange, ED has the advantage of operating as a continuous process, and does not need any regeneration steps, which generally yield liquid wastes of about the same volume as the solution treated. And unlike thermal processes, such as evaporation or crystallization, ED has the advantage of operating only with solutions, at the same temperature, in such a way that the energy consumption for the formation of a new phase (vapor or solid) is eliminated. Moreover, ED used to partially eliminate undesired species of ions is able to separate ions with electrical charges of the same sign, if these are distinguishable by a specific MEM trilogy.

Principle of Electrodialysis

Description of the Method

In an electrodialysis stack, anion exchange membranes permeable to anions are associated in an alternating pattern with cation exchange membranes: those permeable to cations in a filter press-like assembly [6,11]. In industrial applications, the two kinds of compartments so formed are fed in parallel.

Two electrodes, the anode and the cathode form the two extremities and are used in this assembly only to create a constant electrical field, perpendicular to the membranes, strong enough to cause the ions in solution to migrate.

Membranes are formed from organic compounds with a network of fixed charges. They are permeable to one kind of charged species, the counterion, and do not allow ions of the same charge as the fixed charge, the coions, to pass through.

So cation exchange membranes or, more precisely, membranes permeable to cations allow cations to pass through, but not anions, and, conversely, anion exchange membranes or, more precisely, membranes permeable to anions allow anions to pass through, but not cations.

According to Figure 1, under the direct current potential difference applied to the electrodes, the cations travel in the direction of the electric field and leave compartment d7 by passing through cationic membrane c4, but cannot leave compartment b8, as they encounter anionic membrane a4 which stops them. Likewise, anions migrating in the opposite direction can leave compartment d7 by passing through anionic membrane a3, but cannot leave compartment b6 as they are prevented from doing so by cationic membrane c3. Compartment d7 thus loses

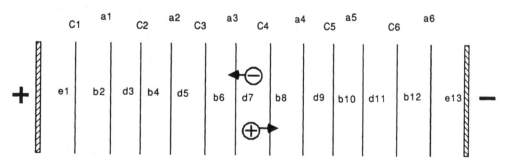

Figure 1   Principle of ion transfer from diluting to concentrating compartments. a, anion exchange membranes; b, concentrate; c, cation exchange membranes; d, diluate; e, electrode rinsing solutions; +, anode; −, cathode; ⊖, anion; ⊕, cation.

dissolved salts; it is known as a dilution compartment "d" in which the treated mixture circulates, while compartments b6 and b8 becomes richer in dissolved matter. These are concentration compartments, or concentrate "b," in which salts are concentrated.

In practice, apart from the two-electrode compartments, "e," an electro-dialyzer consists of a number of these elementary units formed by a membrane permeable to anions and one permeable to cations. In some industrial stacks, this number can be as high as 1000.

From an electrical point of view, the receiver can be considered as a series of resistances. Consequently, most of the voltage drop occurs across the two kinds of solutions and in the two kinds of membranes, the voltage drop across the electrodes and electrode compartments being negligible.

The electric field is perpendicular to the membrane plane.

Generally, solutions circulate through the stack, along the membranes in parallel. Two modes of circulation are in common use:

1. The tortuous path method, used by Ionics Inc. [52], in the USA
2. The flow sheet pattern, used by Corning EIVS [53] and Eurodia in France, Asahi Chemical [48], Asahi Glass [49], and Tokuyama Soda [54] in Japan and in Russian stacks.

Operating Modes

Electrodialysis can be carried out as either a continuous or a batch process. According to the scope of application, either one of these modes is used, or any arrangement, such as "feed and bleed."

In a continuous process, solutions, diluate, and concentrate pass through the stack only once. As in general all the operating conditions of the electrodialyzer are imposed only once, the quantities of material transferred from diluate to concentrate are virtually constant (e.g., see section "Fluxes of Matter and Electricity"). As a consequence, any change in flux or concentration of the feed solution acts on the product.

This difficulty can be overcome by recycling part of the solution. In this way, according to the quantities recycled, fluxes and concentrations fluctuate less and between narrower limits than in the continuous mode.

However, in the recycle mode, the dilution of the solution increases the energy consumption necessary to transfer ions from the diluate to the concentrating solution.

In a batch process, a given quantity of solution is deionized or concentrated up to a desired value. Once this value is reached, the tank containing the solution is discharged. Generally, the entire dilute solution obtained and only part of the concentrated solution are exhausted. With permanent control of the demineralization or concentration level, it is always possible to reach the desired concentration.

If only one part of a given volume of a solution is demineralized in such a way, by mixing the remaining part with the treated part, it is possible to adjust the final concentration of the whole. It generally appears that this mixing may be more economical than a continuous process treating the entire solution.

General Considerations

Main Characteristics

The electrodialysis process has the following characteristics:

1. The concentrate solution can attain 20 wt % or greater if special conditions exist.
2. The concentration of the dilute solution can be reduced to about 100–200 ppm, but must be high enough to give sufficient electrical conductivity.
3. The concentration ratio of concentrated to dilute solutions can be as high as 100.
4. Organic compounds can be separated from electrolytes.
5. The technique is commonly applied for desalination or concentration of neutral strong electrolytes. For example, in salt-manufacturing plants, seawater is concentrated by electrodialysis up to 200 g/liter with an energy consumption of 150 kWh per ton of salt. A special application is the concentration of strong acids, commonly up to 2N. The latter limitation is due to the exchange capacity of membranes currently in use. However, some anion exchange membranes are being disclosed which will allow concentrations to more than 4N, and even 6N.
6. The flux of ions from the dilute solution toward the concentrate is always accompanied by a flow of a certain amount of water. This water transport is mainly caused by ion hydration and electroosmosis, but also by gradients of pH, of hydrostatic and osmotic pressures, and of temperature (see Section "Operating Conditions").

Restrictions

Electrodialysis is subject to the following restrictions:

1. Aqueous solutions must be used. However, contamination with small amounts of organic solvents is tolerable.
2. Care should be taken to localize precipitation of hydroxides or insoluble salts.
3. The pH of the solution for normal electrodialysis should preferably be near-neutral to ensure long membrane life.
4. Strong oxidants should be excluded.
5. The suspended solid content should preferably be less than 1 ppm, and the particle size should not exceed 5 μm.
6. Organic electrolytes of high molecular weight (e.g., detergents or brighteners) should preferably be excluded.

7. The solution temperature should be less than 60°C, and preferably below 40°C. Some attempts are being made to develop high-temperature ED processes, which are limited by membrane materials.

Activated charcoal is sometimes applied during pretreatment in order to eliminate contaminating detergents or brighteners. However, other pretreatment methods may also be applied, such as the use of adsorbing resins. In some cases, pretreatment with a sand filter is used in order to eliminate the suspended solid content, but in most cases a cartridge filter with a pore diameter of about 2 μm is generally sufficient.

Location of Electrodialysis in a Process

Taking into account the fact that ED works well for medium concentration levels and not for removing trace amounts, it should preferably be integrated into a preventative recovery system rather than used as an effluent treatment process. So generally it is preferable to employ ED early in the process.

Before the Main Operation

Located before a principal operation, ED acts as a pretreatment step which brings ionic concentrations to an optimal level for subsequent treatment. In this case, ED can be used either to concentrate desired ions or to eliminate undesired ions before entering the main process step. The latter could be a reactor, a fermenter, a cell, a settler, etc. For example, ED is widely used in conjunction with electrochemical processes, such as metal dissolution or deposition (sections "Application to Hydrometallurgy" and "Application to Surface Treatments"), or with crystallization of salts (sections "Purpose of Electrodialysis" and "Consequences"), or with protein concentration of cheese whey [12].

After the Main Operation

Located after the main reaction operation, ED acts as a cleanup operation, intended to remove some ions from an outgoing flux, as well as to minimize the consumption of raw materials, rather than to reduce pollution.

This mode has the disadvantage of reducing the number of criteria for optimization, i.e., all the parameters concerned with the control of the main operation are excluded from consideration.

Integrated with the Main Operation

Integrated with the principal operation, ED can be used to keep the yield of one or more ions at an optimal value. ED can be used to extract undesired ions from the reactor, thereby achieving a permanent selective and adjustable reactor purge.

Used in another way, ED can introduce some desirable ions into the reactor thereby continuously optimizing the reactor feed. In this case, typically the feed comes from another point in the process.

THEORY OF CONCENTRATION BY ELECTRODIALYSIS

Fluxes of Matter and Electricity

Concentration Polarization

As indicated previously, when a sufficiently high electric field is applied to a membrane stack, the transport of ionic species occurs from the dilute mixture to the concentrate.

But when a counterion is transported from a dilute stream to a concentrate, it is depleted at the membrane-solution interface in the dilute stream, as it is not replaced at this point at the same rate at which it is removed through the membrane. Momentarily, the coion accumulates at the membrane-solution interface but, in order to maintain electroneutrality, the accumulated coion must migrate away instantaneously from the interface into the bulk solution of the dilute stream. Thus, the total ionic concentration in the dilute compartment d is depleted at both interfaces between membrane and solution, as indicated in Figure 2.

On the other hand, the total ionic concentration in the concentrate compartment b accumulates at both interfaces for analogous reasons. These effects occur in a thin zone, namely, the concentration polarization layer (cpl) which is 1–200 μm thick. As such primary concentration polarization occurs, the voltage drop in the dilute stream attains considerable magnitude, since the ohmic resistance of an electrolyte solution is inversely proportional to its ionic content. Consequently, net ionic transport is severely retarded for a given potential. This effect is typically

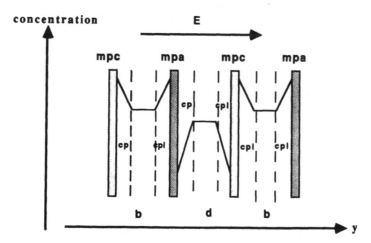

Figure 2   Concentration profiles in a cell pair. E, electric field; b, brine; d, diluate; cpl, concentration polarization layer; mpa, membrane permeable (mainly) to anions; mpc, membrane permeable (mainly) to cations.

reduced by increasing the tangential velocity of streams along the membranes using one of the two modes of circulation previously described in the section "Description of the Method" in order to create turbulent flow. But, if two ions of the same polarity compete in the cpl, the phenomenon becomes quite different (e.g., see section "Cyanide Baths").

Basic Relations

In order to design or operate any ED process it is necessary to have a good understanding of the transport phenomena that occur in a stack. Thus, it is necessary to relate the external electric potential applied at a given concentration distribution across the stack with the current density and the fluxes of ions and molecules. In the first step we consider only transport directly related to the main driving force, i.e., electric field. This affects only transport of ions and, consequently, electricity.

Assuming that the process is running at a uniform temperature and at constant pressure, without convection, the flux density J_i of ion i results from the local difference in concentrations and from the electric field. In the absence of all the other phenomena described below (section "Operating Conditions"), it assumes the following form, known as the Nernst–Einstein equation:

$$J_i = -D_i C_i \, \text{grad} \, (\ln a_i) - z_i C_i D_i \frac{F}{RT} \, \text{grad} \, \phi \tag{1}$$

Conservation of charge yields the electric current density j:

$$j = \sum_i z_i F J_i \tag{2}$$

Combining flux Eq. 1 and charge conservation Eq. 2 gives the electric field, $\text{grad} \, \phi$, which is the same for all ionic species present at the same point in the solution.

Substituting the gradient of the electric potential ϕ such as is obtained in the flux Eq. 1, the expression for the ionic flux J_i can be written as follows:

$$J_i = -D_i C_i \, \text{grad} \, (\ln a_i) - \sum_i \frac{t_i}{z_i} \, \text{grad} \, (\ln a_i) + \frac{t_i}{z_i F} j \tag{3}$$

where t_i is the transport number of species i, that is, the fraction of electric current j transported by species i. If solutions are not too concentrated, the first two terms on the right-hand side of the flux equation depend only on the activity a_i of the ion i alone. Thus, the flux equation can be simplified as follows:

$$J_i = -D_i C_i K_m \, \text{grad} \, (\ln a_i) + \frac{t_i}{z_i F} \tag{4}$$

This equation shows that the motion of an ionic species is always the sum of a diffusion flux and a migration flux. But, in this equation, the coefficient K_m is equal to unity in the case of a solution phase and is very low within a membrane phase [13].

At a steady state, using the expression of the flux density through the membrane (i.e., $K_m = 0$) and through the solution phase (i.e., $K_m = 1$), one obtains:

$$-D_i C_i K_m \, grad \, (ln \, a_i) + \frac{t_i}{z_i F} j = \frac{t_{im}}{z_i F} j \tag{5}$$

In an electrodialysis stack, the flux is perpendicular to the surface of the flat membranes. Furthermore, the integration of Eq. 4 becomes simple if we suppose that the activity a_i of species is equal to the concentration. This yields a relationship between the concentration polarization layer thickness ∂, the electrical current density j, the bulk concentration C_{il} in the liquid solution, and the concentration C_{in} near the membrane for a counterion:

$$C_{in} = C_{il} - (t_{im} - t_i) \frac{j}{z_i} \frac{\partial}{FD_i} \tag{6}$$

When the concentration of the counterion near the membrane becomes zero (i.e., $C_{in} = 0$), the electric current attains its critical value, the limiting current density j_{crit}, expressed as:

$$\dot{j}_{crit} = z_i \frac{F}{t_{im} - t_i} \frac{D_i}{\partial} C_{il} \tag{7}$$

From this equation, it is clear that the critical current appears initially in the dilute stream, and that it increases with the concentration C_{il} of the diluate and the diffusion coefficient D_i. Moreover, its value also increases as the difference in the two transport numbers, $t_{im} - t_i$, and as the concentration polarization layer thickness ∂ decreases. For a given membrane solution, $t_{im} - t_i$ is fixed; so the effect of the thickness ∂ is very important.

The term j_{crit}/C_{il}, called the polarization parameter by Cowan and Brown [14], is nearly independent of the concentration but is related to the velocity and consequently to the Reynolds number, since the thickness ∂ of the concentration polarization layer, which is thinner than the classical hydrodynamic limit layer, is a function of these variables.

But when the ionic concentration at the interface falls to zero, the electric current is still transported by ions arising from the ionic dissociation of the solvent, e.g., hydrogen (hydronium) and hydroxyl ions in the case of aqueous solutions.

The ions so formed will compete with the original counterions of the salt in the transport through the membrane. This reduces the transport of ionic species from the diluate to the concentrate. This also modifies the ionic content of the solution (e.g., by modifying pH if ions remain free), or this modifies ionic equilibria. As a consequence, precipitation can occur near or in the membranes. Therefore, operating conditions are generally chosen such that the current density used is only about three-quarters of the critical current or less.

The Material Balance

At the outlet of each compartment, the net result of the ion flux is either demineralization or concentration, according to the compartment under consideration.

The quantity of salt transferred from one stream to another is directly related to the current density j and the effective area A of each cell pair, the total number of cell pairs being N. As a mole of electricity passes through a unit cell, it transports a given amount of salt.

In industrial cell and stack designs, it is safe to assume that the total ionic flux that flows through adjacent compartments is the same over the whole stack. According to this, the quantity of electricity acts each time it passes through a cell pair. Taking into account the current efficiency η, the number of moles of z:z salt theoretically transported over a time $\Delta\Theta$, for a constant current density j, we have:

$$\Delta n = \eta u \frac{NAj}{zF} \Delta\Theta \tag{8}$$

If ED is operated in a batch mode over a time $\Delta\Theta$, the quantity of matter transferred from the demineralizing volume V_d to the concentrating volume V_b induces a variation in concentrations C_d and C_b of diluate d and concentrate b respectively, such that:

$$\Delta n = -\Delta(V_d C_d) = \Delta(V_b C_b) \tag{9}$$

The two equalities, Eqs. 8 and 9 contain eight variables. Thus, it is necessary to fix six of them in order to solve the material balance. Frequently, the volume of the solution to be treated and its variation in concentration are fixed by local considerations. In the case of a suitable ED stack, the product NA is fixed, and the current density is obtained from the polarization parameter as previously shown. Consequently, one can choose the value of only two variables among the volume of concentrate—its variation in concentration and the duration of treatment. Flow rates can be used instead of volumes, but in this case one has to choose only one variable between flow rate of concentrate and change in its concentration.

The choice of the last variable is very important for calculations. But it is also very important to notice that in any study one cannot choose all the variables freely.

On the other hand, the previous equations are only good approximations as they do not take into account the change in volumes of solutions, diluate, and concentrate, and as they include only transport related primarily to the electric field.

Operating Conditions

Background

When a solution b containing a salt is concentrated by ED, the concentration C_{ib} of an ion i is defined at any time by:

$$C_{ib} = \frac{n_{ib}}{V_b} \tag{10}$$

Under steady-state conditions, the number of moles n_{ib} at the outlet of the concentration compartment is equal to the number n_{ib}^0 of moles at the inlet, increased by the number of moles Δn_{ib} transferred from the solution being diluted to the solution being concentrated [15,16].

The final volume V_b is equal to the initial volume V_b^0 plus the volume which passes from the diluate to the concentrate. In general, in addition to the low volume of the stripped ions, the volume of a certain number of solvent molecules must be added. With aqueous solutions, this type of water transport limits the maximum concentration which can be attained.

Solvent transport can be analyzed in several ways. In the following, the solvent will be water, but the reasoning holds true for other solvents as well.

In each kind of membrane is considered individually, water transport can be defined by an electroosmotic transport number [9], which in turn is defined as being equal to the volume of water transported by the charge of 1 mole of electricity, i.e., 1 faraday. Determination of the electroosmotic transport number is relatively simple for given conditions [17,18]. However, the theoretical expression of the electroosmotic coefficient shows that numerous conditions must be respected [13]. Certain authors [19] have proposed that the water transport number should be related to that of the counterion. In fact, this comes down to associating the number of moles of ions transferred. It is clear that at this stage it is indispensable to take the hydration number of the ions into account. But the experimental determination of hydration number of an ion is often difficult and depends on the method of measurement used [17,20].

In some cases, the electroosmotic flux itself has been used to determine the water transport number [21]. The water transport number allows a correction to be made for the apparent value of the ion transport number when there is a flow of water [22].

Considering that the process uses pairs of membranes, determinations involving an isolated membrane are often problematic to apply to pairs of membranes.

Overall, the water transported from the diluate to the concentrate is considered as contributing to the reduction of the efficiency of the separation system. This decrease in efficiency is sometimes taken into account by introducing an additional term reducing the coulomb yield, i.e., the overall proportionality constant between the density of salt flux and the density of the electric current [6].

Also, it can be useful to link the transported water to the hydrodynamic [23] or electric [24] conditions.

One of the most interesting ways is to make use of the relationships used in the thermodynamics of irreversible processes relating the flux to the true forces acting on species, ions, or molecules [25].

Final Volume of Concentrate

The volume V_b at the outlet of the concentration circuit of the electrodialyzer is equal to the initial volume V_b^0 plus the volume transferred from the diluate to the concentrate but corrected for the variation of volume due to the change of concentration:

$$V_b = \pi V_b^0 + \Delta V_b \tag{11}$$

The correction factor π, due to the solvation of the solute, can be determined directly from measurements of the specific mass depending on the molarity [26]. Generally, the coefficient π is not far from unity.

The volume transferred ΔV_b is due to all the species, ions, or molecules which leave the dilute solution and enter the concentrate. If the density of the transmembrane flux of a species by k, ion or molecule, is represented by J_k and its molar volume by v_K, the volume transferred across area A during a time $\Delta\Theta$ can be written as:

$$\Delta V_b = \sum_k \int_0^A \int_0^{\Delta\Theta} J_k v_k \, dA \, d\Theta \tag{12}$$

If the molar volumes are assumed to be constant at steady state (i.e., for constant flux densities), the previous expression becomes:

$$\Delta V_b = \sum_k J_k v_k A\Theta \tag{13}$$

In the case where the flux densities J_k are directly proportional to the general forces X_1, as done in the thermodynamics of irreversible processes [27], they can be expressed by the classical relationship:

$$J_k = \sum_l L_{kl} X_l \tag{14}$$

Then the overall volume flux J_{vb} is such that

$$\Delta V_b = J_{vb} A \, \Delta\Theta \tag{15}$$

where we have

$$J_{vb} = \sum_l \sum_k L_{kl} v_k X_l \tag{16}$$

Using now the expression of the generalized forces X_1 in the form of the isothermal chemical potential μ for the molecules, and in the form of the isothermal electrochemical potential $\bar{\mu}$ for ions, we obtain a relationship including a term related to the concentrations of the solvent in the diluate and the concentrate, such as the gradient of the osmotic pressure π_w, a term related to the concentrations of the ions of the salt by means of their activities a_s, a term due to the concentrations of ions resulting from the ionization of the solvent, such as the gradient of pH, a term due to the gradient of the hydrostatic pressure P, a term due to the gradient of the potential U applied to each cell pair, and a correction factor ln B. Parameter B is related to the ratio of the applied current to the critical current and to the demineralization factor; it accounts for the real forces acting on species. This last term is necessary to take into account the fact that the concentrations close to the membranes are different from those in the bulk of the solution [25]:

$$J_{vb} = k_1 \operatorname{grad} \pi_w + k_2 \operatorname{grad} \ln a_s + k_3 \operatorname{grad} pH + k_4 \operatorname{grad} P +$$
$$k_5 \operatorname{grad} U + k_6 \ln B \tag{17}$$

It should be noted that the six coefficients k_1–k_6 cannot be associated with phenomenological coefficients as defined by irreversible thermodynamics.

If in the calculations the electric potential taken into account is the overall potential applied to the stack of membranes, which in industrial electrodialyzers is not very different from that including the potential drops on electrodes, one may add a correction factor equal to $-k_5 \ln N$, where N is the total number of cell pairs.

As an example, Table 8 gives the values of the six coefficients for the concentration of zinc sulfate solutions in a laboratory tortuous path electrodialyzer using heterogeneous membranes ARP and CRP of the Rhône Poulenc Techsep Society, with operating conditions such that the electric current is lower than the critical current.

Table 9 gives the contribution of each driving force to the total volume flux. It is noted that the major contributions to the transport of water from diluate to concentrate are due to the gradients of pH and potential in the positive direction, and to the term related to the concentration of salt in the negative direction. The minor contributions are those due to pressure gradient, osmostic or hydrostatic, and the lowest is due to the correction factor ln B.

Table 8 Values of Pseudophenomenological Coefficients for Total Flux in the Case of Zinc Sulfate Concentration

$10^{15}k_1$ m·sec^{-1}·Pa^{-1}	$10^9 k_2$ m·sec^{-1}	$10^9 k_3$ m·sec^{-1}	$10^{12}k_4$ m·sec^{-1}·Pa^{-1}	$10^9 k_5$ m·sec^{-1}·V^{-1}	$10^9 k_6$ m·sec^{-1}
-7.923	-220.90	407.88	1.139	3.804	18.047

Source: Data from Ref. 25.

Table 9 Contribution to Total Flux Density of Each Driving Force, in %, Positive from Diluate to Concentrate

Run	Due to:	πW	$\ln a_s$	pH	P	U	$\ln B$
1		−5.92	−98.82	137.03	6.75	57.37	3.59
2		−6.27	−93.76	139.20	7.41	48.14	5.27
3		−5.81	−85.06	130.98	13.94	40.49	5.45
4		−1.61	−55.38	105.89	1.01	48.78	1.30
5		−1.55	−36.27	105.00	0.33	29.90	2.58
6		−5.67	−45.19	97.20	6.01	30.42	17.23

Electrolyte Ion Flux

The number of moles n_{ib} of ion i at the outlet of the electrodialyzer, in the concentrate stream, is equal to the number n_{ib}^0 at the inlet plus the number of moles Δn_{ib} transferred from the solution being diluted to the concentrate:

$$n_{ib} = n_{ib}^0 + \Delta n_{ib} \tag{18}$$

If J_i represents the transmembrane flux density, the number of moles transported across surface area A during the time $\Delta\Theta$ is:

$$\Delta n_{ib} = \int_0^A \int_0^{\Delta\Theta} J_i \, dA \, d\Theta \tag{19}$$

At steady state, the density J_i of the transmembrane flux is constant and thus:

$$\Delta n_{ib} = J_i A\Theta \tag{20}$$

In the case where the flux densities J_i are proportional to the general forces X_I (i.e., to the gradient of an extensive variable, as in irreversible thermodynamic), they are expressed by the following relationship [28]:

$$J_i = \sum_{i=1}^{k} L_{il}X_l \tag{21}$$

J_i and X_l being the vectors.

By distinguishing the isothermal contributions and in the absence of chemical reactions, the fluid being perfect, the general force X_l for the ions is the electrochemical potential $\bar{\mu}$ which involves, as well the chemical potential gradient μ, the gradient of the electric potential ϕ, so that

$$X_l = (z_l F \, \text{grad} \, \phi + \text{grad} \, \mu_l)_T \tag{22}$$

the charge number z_l being an algebraic magnitude [29].

Finally, as in the case of the overall flux, the flux of ions is given by a relationship with six pseudophenomenological coefficients:

Table 10 Values of Pseudophenomenological Coefficients for Total Volumic
Ionic Flux in the Case of Zinc Sulfate Concentration

Coefficient					
$10^{15}k_{i1}$ $m \cdot sec^{-1} \cdot Pa^{-1}$	$10^9 k_{i2}$ $m \cdot sec^{-1}$	$10^9 k_{i3}$ $m \cdot sec^{-1}$	$10^{12} k_{i4}$ $m \cdot sec^{-1} \cdot Pa^{-1}$	$10^9 k_{i5}$ $m \cdot sec^{-1} \cdot V^{-1}$	$10^9 k_{i6}$ $m \cdot sec^{-1}$
0.315	−5.06	7.48	96.92	0.969	0.969

$$J_i = k_{i1} \operatorname{grad} \pi_w + k_{i2} \operatorname{grad} \ln a_s + k_{i3} \operatorname{grad} pH + k_{i4} \operatorname{grad} P +$$
$$k_{i5} \operatorname{grad} U + k_{i6} \ln B \qquad (23)$$

As an example, Table 10 gives the values of these six coefficients for the concentration of zinc sulfate solutions, in a laboratory tortuous path electrodialyzer using heterogeneous membranes ARP and CRP of the Rhone Poulenc Techsep Society, with operating conditions such that the electric current is lower than the critical current.

Table 11 gives the contribution of each driving force to the total volumetric ionic flux, which is about 30 times lower than the total volumetric flux given in Table 9. It is noted that the major contributions to the transfer of salt from diluate to concentrate are due to the gradients of pH and potential in the positive direction , and to the term related to the concentration of salt in the negative direction. The minor contributions are those of the gradient in osmotic or hydrostatic pressures.

Consequences

High Concentration of Concentrate

Taking into account the expressions of the variations in volume ΔV_b and in number of moles of ions Δn_{ib}, the concentration C_{ib} of an ion i in concentrate b is equal at any time to:

Table 11 Contribution to Total Salt Flux Density of Each Driving
Force (%), Positive from Diluate to Concentrate

Run	Due to	πW	$\ln a_s$	pH	P	U	$\ln B$
1		8.69	−84.27	93.66	27.31	47.45	7.17
2		8.42	−72.44	86.10	27.07	41.34	9.51
3		6.52	−55.34	68.26	43.04	29.30	8.26
4		2.84	−57.70	90.10	4.98	56.58	3.19
5		2.93	−39.70	92.16	1.67	36.36	6.58
6		6.76	−31.09	53.57	19.62	23.28	27.86

$$C_{ib} = \frac{n_{ib}^0 + J_i A \Delta\Theta}{\pi V_b^0 + J_{vb} A \Delta\Theta} \tag{24}$$

Then, the development of this expression shows that for an initial concentration C_{ib}^0 the variation of the concentration ΔC_{ib} can be written in the form:

$$\Delta C_{ib} = C_{ib}^0 \frac{k_{1x} + k_{2x}x}{k_{3x} + k_{4x}x} \tag{25}$$

where x is one of the following variables: total area A, time $\Delta\Theta$, or any gradient, among osmotic pressure, hydrostatic pressure, pH, logarithm of salt activity, and electric potential.

The total working area A is equal to the active part of the whole surface of cell pair membrane installed in the stack through which ions and molecules are transferred from one compartment to another, times the number N of cell pairs:

$$A = Na \tag{26}$$

On the other hand, it is interesting to note that the gradient of the applied potential U is related to the electric current density j by Ohm's law when operating conditions are lower than critical [30,31]. As a consequence, x may be the current density j passing through the stack instead of grad U. The coefficients k_{1x}, k_{2x}, k_{3x}, and k_{4x} are the constants associated with the chosen x variable.

Relationship 25 gives the best operating conditions for obtaining a given change in concentration ΔC_b. It indicates that for a given pair of ion exchange membranes, for which the transport numbers are known, the increase in concentration of the concentrate stream is dependent only on the operating conditions.

It appears that the maximum variation in concentration ΔC_b, and consequently the upper limit of the concentrate C_b, is reached when the following condition is met:

$$k_1 k_4 = k_2 k_3 \tag{27}$$

For example, in the case where x is the time ($x \equiv \Delta\Theta$), Eq. 25 can be linearized in terms of the inverse of $\Delta\Theta$:

$$\frac{C_{ib}^0}{\Delta C_{ib}} = \frac{1}{p - q}\left(1 + \frac{1}{\Delta\Theta}\right) \tag{28}$$

This linear relationship has been verified for several kinds of devices for zinc sulfate solutions and for various operating modes [25], as shown in Figure 3.

In practice, results show that is possible using ED to concentrate a solution more than 20 times, and up to 100 sometimes, if the operating conditions can be completely controlled.

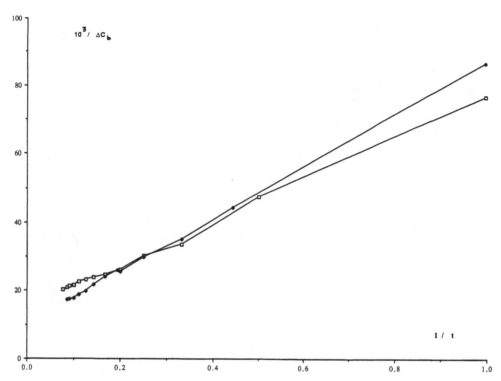

Figure 3 Example of linearization of the increase ΔC_b of concentration of concentrate with time t. Example of zinc sulfate solutions [25]. C_b in g Zn/liter: ◆: $C_b^0 = 6$, $C_b^f = 55.5$; □: $C_b^0 = 40$, $C_b^f = 98$ g/liter.

Scale Formation

From the previous discussion, it appears that the change of concentration in the concentrate is mainly a function of the operating conditions.

The theoretical approach described in the section "Operating Conditions" concerns only fully ionized salts. But, either for the partially or fully ionized component, the trends remain the same. This is the case, for example, for strong acids treated by ED, such as hydrochlorhydric, hydrofluoric, sulfuric, nitric acids, or weak acids, such as acetic, cyanuric, and chromic acids.

Moreover, the previous discussion deals with a single salt in solution. However, in many cases the aqueous solution contains many kind of salts, and/or acids or bases mixed together. So there is always a competition between the different kinds of ions in passing through the membranes as counterions, or in some cases as coion and counterion, under the influence of the applied driving forces and those which appear. So, resulting either from the electrical potential gradient or the concentration and/or the pH gradient, or from the hydraulic or the osmotic pressure, or any other gradient, the fluxes of ions are always unequal. From Eq. 3, then, one can conclude that transport values and diffusion coefficients differ from one ion to another. In practice, this yields very different values of the pseudo-phenomenological coefficients k_i of Eq. 23. For example, the flux of ions from the ionic decomposition of water (proton and hydroxyl ion) is generally greater than the flux of other ions. This induces modifications in the pH of the diluate or concentrate. In some cases, the depletion of protons is such that the value of the ionic product is reached for a given salt, and precipitation occurs. For example, during water desalination, great care is always taken to avoid the scaling of membranes by formation of calcium sulfate precipitates.

The same phenomena can occur with any complexing agent, such as, for example, cyanide ions (see section "Application to Surface Treatments"). However, these precipitates rarely appear in the middle of a compartment, but rather on the surface or inside a membrane. Thus, there are typically layers coating the surface or deposits plugging the pores of the membrane. Since the conductivity of a solid salt is lower than the conductivity of its dissolved ions, the electrical resistance of the membrane increases in the region where the precipitate is located. As a consequence, the electrical current decreases, and the overall efficiency of the stack decreases. In some cases, the membranes become useless and must be removed.

To avoid these phenomena, one must know the physicochemical behavior of the solutions treated and the streams from the electrodialyzer as concentrate and even as diluate.

High Recovery Rate

The concept of polarity reversal has created new possibilities for electrodialysis. In electrodialysis reversal (EDR) the direction of current and hence of ion fluxes are reversed periodically, thus controlling the scaling of membranes. With electrical polarity reversal, continuous operation without expensive chemical pretreatment is possible, even in the region of supersaturation of scaling components [32]. Naturally, the concentrate and the diluate compartments and the polarity of electrodes must be reversed, which requires a suitable design and additional piping.

Although deposit formation in or on membranes can be avoided by polarity reversal, scaling may still occur in valves, piping, tanks, etc. Such scaling and blocking can be avoided by controlled crystallization. In order to achieve this, an electrodialysis stack, with polarity reversal, can be associated with a crystallizer, as indicated in Figure 4.

Provided that the crystallizer is designed properly with respect to residence time, supersaturation of retentate can be eliminated. The crystal-free solution is recycled to the electrodialyzer and the crystals are continuously extracted at the

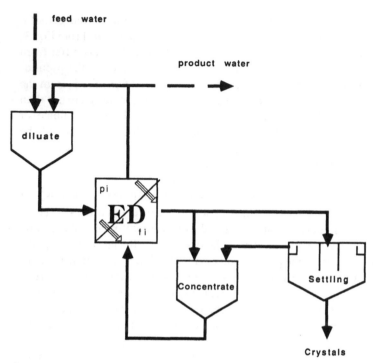

Figure 4   Continuous extraction of crystals from concentrate.

bottom of the crystallizer. Pilot plant experiments have shown that 99% of the salt contained in the feedwater can be extracted. In this case, the feed solution contained 284.4 mg/liter of calcium sulfate, and the concentrate attained concentrations as high as 27 g of total dissolved salt (TDS) per liter [33].

Finally, even though crystal deposition or scale formation is an important problem for electrodialysis, it can be controlled by adjusting the operating conditions depending on the physicochemical properties of the solutions treated, feed, and stack effluent.

ECONOMIC ASPECTS

Basics of the Balances

Before the construction of any plant for waste concentration by ED or in running an existing plant, it is always necessary to have good estimates of the cost of the operation. But because the concentration of waste by ED is always associated with its demineralization, as a result, the process produces both water with a low mineral content and a concentrated solution, both suitable for reuse.

The economic aspect of the operation is simply how much can be saved, and this shall be established on a fixed basis. At this point, many types of bases are possible. An annualized basis is typically used, since it is the normal duration of an economic balance in any firm.

Technical personnel prefer the choice of material used, produced, or wasted. Indeed, there is no difficulty in translating from the technical basis to the financial, as the annual amount of material is always known. But a material basis allows comparisons to be made between plants.

So in the following, the basis chosen for the economic analysis of the concentration of waste by ED is a unit of material saved, i.e., in international units, 1 kg of matter. But this quantity can also be expressed as pure metal or as a salt. Since in an ED process metal exists only as ions in solution, it appears better to use, as a basis, the unit quantity of principal salt in the effluent treated and, in some cases, the unit volume treated.

But in addition to direct savings, some indirect savings occur which only become apparent after sustained operation. These occur particularly when ED can be converted from a batch to a continuous process. As a consequence, the quality of the products increases and losses decrease. But these savings can be calculated only by comparison of the efficiency of the plant, before and after the installation of the ED process.

In the following, only the direct savings are discussed.

Investment Costs

Before the determination of the unit cost of ED treatment of 1 kg of material saved, it is necessary to know the overall cost [34]. This must be calculated for the total period t during which the plant runs. The overall cost, CI, includes the investment costs, CA, all the annual operating costs, COA, and all the annual charges, CFF:

$$CI = CA + \sum_t COA + \sum_t CFF \qquad (29)$$

Without significant error, it is possible to assume that the annual costs and the annual charges are constant during the entire working period of the installation. So the overall cost can be expressed as:

$$CI = CA + t \cdot COA + t \cdot CFF \qquad (30)$$

On the other hand, if we assume that the annual quantity of material saved is constant over the entire period t, and that is rate is M^* (expressed in kg per year), the total quantity of material saved by the plant during its life is M^*t. With these assumptions, the unit overall cost, CUI (i.e., the cost of the unit quantity of material saved) is:

$$\text{CUI} = \frac{\text{CI}}{\text{M*}\cdot\text{t}} = \frac{\text{CA}}{\text{M*}\cdot\text{t}} + \frac{\text{COA}}{\text{M*}\cdot\text{t}} + \frac{\text{CFF}}{\text{M*}\cdot\text{t}} \tag{31}$$

The investment cost is generally proportional to the effective surface area A of membrane used. If the price of the unit area of membrane is v_{AM}, the investment cost can be written as:

$$\text{CA} = k_A \cdot A \cdot v_{AM} \tag{32}$$

where k_A is a proportionality factor including the costs of both kinds of membranes, CAM; of frames, CAT; of the press, CAB; of the continuous electric power supply station, CAA; of motor pumps, CAP; of buildings, CAH; of land, CAS; and of erection and assembling, CAE. In some cases it is necessary to add the cost of the pretreatment station, filter, ion exchange resins or others, CAF. All these costs must be written off over the period t.

Moreover, it is logical to include in the investment cost the cost of the regular phasing out of the membranes. That is, if membranes are not regularly replaced, the plant cannot run in its nominal condition. In fact, during the working period t of the plant it is invariably necessary to change some membranes. However, this does not occur annually, but only two or three times during the life of the plant. So the cost of this replacement, CAR, is two or three times the cost of the initial membranes.

Taking all the preceding into account, the total investment cost can be expressed as:

$$\text{CA} = \text{CAM} + \text{CAT} + \text{CAB} + \text{CAA} + \text{CAP} + \text{CAH} + \text{CAS} +$$
$$\text{CAE} + \text{CAF} + \text{CAR} \tag{33}$$

As a consequence, the proportionality factor k_A is greater than unity, and typically varies from 10 to 20, according to local conditions.

Finally, the unit investment cost is expressed as

$$\text{CUAI} = \frac{\text{CA}}{\text{M*}\cdot\text{t}} \tag{34}$$

Annual Costs

The annual costs include the operating costs, COA, and the fixed annual costs, CFF. The CFF depend on the internal conditions of the firm. They do not include the financial charges of the plant, which are taken into account in the investment cost, but they cover part of the total charges of the firm and the working costs directly related to the ED process. Generally, in an ED plant used for waste treatment, there are no costs, or very low costs, for manpower directly concerned with the stacks, since the installation runs unattended. But in many plants, some personnel working in other areas are used to monitor the process. As a consequence, this cost is omitted or roughly estimated as a part of the investment cost. Thus, the fixed unit cost is:

$$CUFL = \frac{CUFF}{M^* \cdot t} = k_L \cdot CUAI \tag{35}$$

The operating costs, COA, include the cost of transfer of ions from the diluting compartment to the concentrating compartment, CTI, directly related to the current density j. They also include the cost of pumping CFP of the three solutions—diluate, concentrate, and electrolytes—through the stack. Moreover, in some cases it is necessary to add the cost of pre- or posttreatment, CFT, filter changes, resins changes, etc., and of cleaning the stack, CFN. Thus, the annual operating cost, COA, can be expressed as:

$$COA = CTI + CFP + CFT + CFN \tag{36}$$

Finally, the unit operating cost, CUFO (i.e., the operating cost for 1 kg of material saved) includes four terms:

$$CUFO = \frac{t \cdot COA}{t \cdot M^*} = \frac{CTI}{M^*} + \frac{CFP}{M^*} + \frac{CFT}{M^*} + \frac{CFN}{M^*} \tag{37}$$

which can also be written as:

$$CUFO = CUTI + CUFP + CUFT + CUFN \tag{38}$$

The energy consumption per unit of material saved, E, can generally be easily directly measured to take into account all the coulomb efficiencies [5]. As a result, if the unit price of energy, v_{AE}, is known, the unit cost for the transfer of ions through the membranes is:

$$CUTI = E \cdot v_{AE} \tag{39}$$

As indicated previously, some analysts add the cost of membrane change, CAR, but in this case, this cost must not be included in the investment costs, CA. The unit contribution of membrane replacement is then:

$$CUAR = \frac{CAR}{M^* \cdot t} \tag{40}$$

Costs of Products Treated

The purpose of waste treatment by ED is to change the composition of the effluent and to obtain two different streams with two different concentrations. So, it is necessary to take into account the price of metals or metalloids saved in the concentrated solution. For a unit of material saved, this price is the price of the component in its salt form, CUVM.

However, it is also of interest to take into account the price of the low-mineral-content water, which is generally recycled in the plant. In some cases, a complementary treatment by ion exchange resins yields pure demineralized water. So the price of the purified water obtained as diluate, per unit of quantity of material saved, CUVD, depends greatly on the possibilities of reusing it within the plant.

Moreover, in some cases, the wasted effluent treated by ED has an associated cost, due to taxes or transportation fees, for example. Thus, the price per unit of material saved, CUVE, can be negative or positive, according to local conditions.

Finally, the money saved per unit quantity of material treated by ED is:

$$CUV = CUVM + CUVD + CUVE \tag{41}$$

Overall Balance—Turnover

The overall balance includes all the unit costs presented above.

The net profit per unit of material saved is then the difference between the cost of operation and the amount earned by using ED. Consequently,

$$\text{Net profit} = CUV - CUI \tag{42}$$

which is the net profit per unit of material saved, for a period of time t, during which the annual rate of material recovered is M*.

In some cases, it is interesting to determine the turnover time t_0, i.e., the period necessary to recoup the investment costs. Using the previous relations, this value may be calculated by equating the cost of investments, CA, to the difference between earnings, $CUV \cdot M* \cdot t_0$, and the operating and fixed costs, $(COA + CFF) \cdot t_0$, that is:

$$CA = t_0 \cdot [M* \cdot CUV - (COA + CFF)] \tag{43}$$

For treatment of galvanization effluents, for example, the turnover time may be less than a year.

APPLICATION TO HYDROMETALLURGY

Electrowinning of Metals

In a hydrometallurgy plant, metal is extracted from its naturally occurring ores by an electrochemical reaction, during which the metal cation is reduced on a cathode. In fact, only a small number of metals are manufactured solely by electrowinning [35]. This number is increasing. In the periodic table of the elements, these are mainly solids from Group IV with an atomic number ranging from 22 to 31, i.e., Ti, Cr, Mn, Co, Ni, Cu, Zn, and Ga, which is a liquid, and Cd, In, and Ti. Most of these exist as divalent cations, M^{2+}.

In hydrometallurgy, electrodeposition occurs from a solution containing the metal in an appropriate solvent, which in most cases is water. Indeed, the favorable properties of water as a solvent are well known and include the ability to dissolve salts to relatively high concentrations, and to produce solutions with high conductivities at ambient temperature, especially when acidic.

In general the ore is converted to an acid-soluble form, usually an oxide, by roasting. The product is then leached with an acid and the metal thus obtained is in the form of a dissolved cation, as indicated by:

$$MO_x + 2xH^+ (aq) \rightarrow M^{2x+} (aq) + xH_2O \tag{44}$$

After purification and concentration, electrolysis takes place, during which the cathodic reaction is of the type:

$$M^{2x+} (aq) + 2xe^- \rightarrow M^0 \tag{45}$$

and the anodic reaction is of the type:

$$xH_2O \rightarrow 0.5xO_2 + 2xH^+ (aq) + 2xe^- \tag{46}$$

The acid so produced at the anode is recycled to the leaching bath. The net reaction can be written simply as:

$$MO_x \rightarrow M^0 + 0.5xO_2 \tag{47}$$

Theoretically, there is no overall consumption or production of acid or water.

Normally, the cathode is flat in sheets or it is cylindrical. In some cases, the cathode is in the form of granules, held in suspension in the solution, sometimes as a fluidized bed [36].

Once the metal is obtained as a sheet, powder, granules, etc., it is extracted from the electrodeposition bath. Even when the solution is drained off the metal by capillary effect some acid adheres to it. As a consequence, it is necessary to rinse it. The rinse water contains the solution obtained during electrodeposition. According to the particular situation, the dilute solution obtained must be reconcentrated to recycle the metal and reduce the losses, or its salts must be eliminated to reduce pollution. ED is an efficient means for performing both operations, i.e., concentration of metal and demineralization of the dilute solution.

Zinc Metallurgy

During the electrochemical production of zinc sheets in sulfate baths, the rinse water is usually too dilute to be reintroduced into the deposition tank. Preliminary concentration before recycling can be obtained by ED.

Nevertheless, analysis of the system shows that the values of the four volumes or flows as well as those of the four concentrations involved must be known. For every device, there are two equations, like relation Eq. 9, accounting for matter, one for the dissolved matter, the other for the solvent, and two equations, like relation Eq. 3, describing the flow of material between the dilute solution and the concentrate. It is thus possible to fix only four variables at an arbitrary value; the other four variables result from solving the system of equations. In this way, as indicated in the section "The Material Balance," if two inlet concentrations and one volume are given, only one other choice is possible.

If, for reasons associated with pollution requirements, the output concentration of the dilute solution is fixed, the concentrated solution may not be able to be recycled to the electrolysis bath. On the other hand, if, for reasons of economy of raw material, the output concentration of the concentrated solution is imposed by the electrolysis bath, it is possible that the dilute solution concentration might be above the level allowed for dumping. To satisfy these two conditions, the concentrate is reintroduced into the electrodeposition bath, and the diluate is used as primary rinse water or is included in another circuit, in agreement with the general water balance of the plant.

Two different schemes have been investigated in this latter regard, and the matter, energy, and economic balances have been established [1].

The first flow sheet uses one an ED stack as indicated in Figure 5. The dilute solution circulates according the feed-and-bleed mode, and the concentrate is batch mode. This concentrate is recycled between its own tank and the ED stack until its concentration reaches ~100 g of Zn per liter, suitable for reintroduction

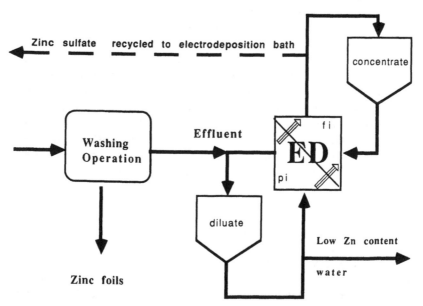

Figure 5   Concentration of zinc sulfate in an effluent by single-stage ED with batch recirculation of concentrate.

into the electrodeposition bath. Using this route, it is possible to save about 98% of the amount of zinc discharged. But during the treatment of over 700 m³/day of an acidic zinc sulfate solution containing 6 g of zinc per liter, the amount of water transferred is on the order of 5% of the total volume treated. For these conditions, the energy consumption is 7.2 kWh/kg zinc recovered.

In the second process, the wash solution is continuously treated by two stages of ED stacks. The effluent still flows according to the feed-and-bleed mode in the first stack, but the concentrate product is sent as diluate into the second ED stack, as indicated in Figure 6. For practical reasons, the quantities and the concentra-

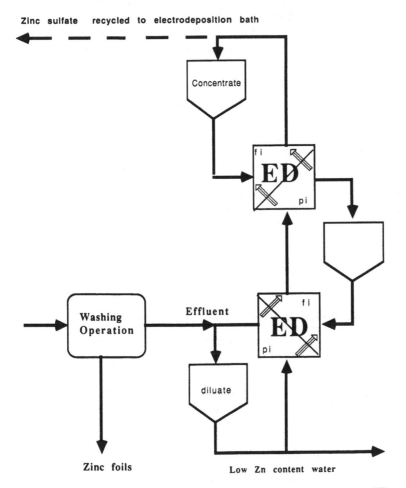

Figure 6 Concentration of zinc sulfate of an effluent by two-stage ED with batch recirculation of concentrate.

Table 12A Treatment of a Zinc Effluent by ED: Specifications

Effluent treated		
Volume	714 m³/day	
Concentration	6 g Zn/liter	
Recovery rate	98%	
Molecular weight	65.4 g/mole Zn	
Electrodialysis stacks (SRTI)		
Membrane spacing	0.5 mm	
Membrane size	0.5 m × 0.5 m × 1 m	
Effective area	70%	
Number of cell pairs	Stack 1:800	Stack 2:1200
Operating conditions	Stack 1	Stack 2
Current density	130 mA/cm²	90 mA/cm²
Energy expenditure	2.8 kWh/kg Zn recovered	2.0 kWh/kg Zn recovered
Tangential flow rate	4.9 cm/sec	4.9 cm/sec
Pressure loss	0.1 bar	0.1 bar
Coulomb efficiency	80%	75%
Coefficients		
Investment	18.5	9.25
Pumps	2.3	2.3
Motor-pump efficiency	0.5	0.5
Ion transfer	2	2
Unit prices		
Membranes	1000 francs/m²	
Electricity	0.15 francs/kWh	
Zinc	16 francs/kg	

Source: Data from Ref. 1.

Table 12B Economic Analysis, in French Francs per kg of Zinc Saved

Cost of	Abbrev.	Stack 1	Stack 2	Both stacks
Investment	CUAI	2.94	4.40	7.34
Pumping	CUFP	0.02	0.03	0.03
Ion transportation	CUTI	0.84	0.60	1.44
Labor (3% of CUAI)	CUFL	0.09	0.13	0.13
Total cost	CUI	3.89	5.16	9.05
Price of zinc purchased				16.00
Net profit, in French francs per kg of zinc saved				6.95

Source: Data from Ref. 1.

tions of the feed and product solutions from the two stacks are the same as in the previous case. However the overall energy consumption is reduced to 4.8 kWh/kg zinc saved, but the water transferred from the diluate to the concentrate is still on the order of 5% of the volume of the effluent treated.

The main specifications of this ED facility for a French plant are given in Table 12A. The economic analysis is given in Table 12B for the case of a plant operating for 15 years, 300 days a year and 24 hours a day. The net profit is about 7 French francs per kg of zinc saved, i.e., more than 8 million French francs per year.

APPLICATION TO SURFACE TREATMENTS

Electrodialysis is generally used for the separation of strong electrolytes. Effluents from galvanization are typical examples of inorganic effluents of this kind. Effluents from galvanization are treated to recover useful dissolved components or for regeneration of water for reuse.

From an economic point of view, metals are recovered according to their local or international cost. Table 13 gives some costs of some metals currently used in surface treatments. From a practical point of view, slightly acidic surface treatment effluents are easier to treat than strongly acidic or basic effluents.

Recovery of Nickel by Electrodialysis

Nickel recovery by means of electrodialysis or ion exchange is widely practiced. The effluent discharged after washing during nickel plating, containing nickel, chloride, and sulfate ions, is treated by ED. As a result, nickel salts are concentrated for recovery while the dilute solution can be reused as a wash solution. The concentration ratio of the concentrated solution to the dilute solution is easily greater than 70.

As shown in Figure 7, the electrodialyzer is equipped with a wash tank in the first stage of a galvanization line having multistage countercurrent washing tanks. Most of the nickel discharged with products from the electrodeposition cell can be recovered and returned to the galvanization bath. In addition, washing efficiency in the later stages is improved, as the solution to be washed from the products is more dilute.

Generally, some organic electrolytes are present in the additives used in the galvanization bath, so small amounts of these inorganic substances are discharged in the effluent from washing. This results in the contamination of the ion exchange membranes. These organic electrolytes must therefore be eliminated at the pretreatment stage for electrodialysis, with active charcoal or with weak electrolytes.

Table 13 Cost of Some Useful Metals in Pounds Sterling/kg at the Bourse
de Commerce de Paris (BCP), or at the New York Commodity Exchange
(COMEX), or at the London Metal Exchange (LME) During Summer 1989

Aluminum	1.09	(LME)	Lead	0.43	(LME)
Cadmium	9.50	(BCP)	Nickel	8.18	(LME)
Chromium	3.13	( )	Platinum	9,690	(COMEX)
Cobalt	12.2	(BCP)	Silver	103	(COMEX)
Copper	1.60	(LME)	Tin	6.27	(BCP)
Gold	7,315	(COMEX)	Zinc	1.11	(LME)

The waste solution discharged after washing, and purified as above, is circulated through the diluting compartments of the electrodialyzer and demineralized until the desired concentration is attained, to the bath mode. The dilute solution is then directly utilized as wash water. After its concentration and pH have been adjusted, the concentrated solution is returned to the galvanization bath.

The quantity of nickel salts transferred from the diluate to the concentrate in the ED stack is determined on the basis of the quantity of nickel discharged from the electrodeposition tank. In general, the recovery of nickel discharged from the galvanization bath is approximately 90% or greater. Thus, the calculation of the desired surface area of membrane used for ED is based on the quantity of nickel treated and on the current density as indicated by Eq. 8.

Nickel salts, such as nickel sulfate and nickel chloride, are hydrolyzed, even in neutral aqueous solutions. As a result, precipitation of nickel hydroxide occurs. So special considerations must be taken regarding the operating conditions for ED.

The main specifications for ED facilities of a Japanese plant are given in Table 14A. In the case of an effluent containing 5–7 g of Ni^{2+} per liter, calculations were performed for a plant operating for 7 years, 300 days a year and 21 hours per day. An economic analysis (Table 14B) indicates a net profit of 425 yen/kg of $NiSO_4$ saved, i.e., about 12 million yen/year.

Recovery of Copper by Electrodialysis

The recovery of copper by electrodialysis is economically viable. However, the electrodialysis baths are basic. Therefore, to prevent metal hydroxide precipitation, cyanide ions are added.

The insertion of an electrodialyzer between the basic electrodeposition cell and the first static rinse tank allows most of the copper cyanide to be recovered. The copper in the effluent is withdrawn from the diluate and returned as concentrate to the electrolysis bath. With these operating conditions, ED is performed with a total recycle of solutions, as in a batch mode, but with constant concentra-

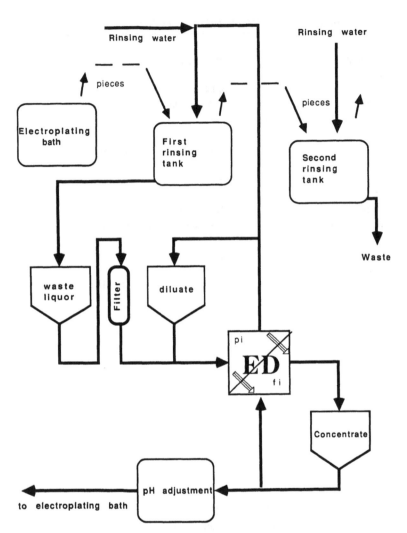

Figure 7   Concentration of nickel sulfate during electroplating.

tions, as in a continuous process. As a consequence, the energy expenditure is lowered, to about 1–2 kWh per kg of copper saved [37].

But during metal transfer, as indicated in the section "Operating Conditions," water passes from the diluate to the concentrate. Indeed, this is not a disadvantage, as the plant generally is in a state of water imbalance. So, as in the case of most surface treatment processes, it is necessary to supply makeup water to compensate

Table 14A   Treatment of a Nickel Effluent by ED: Specifications

Effluent treated	
Nickel in effluent	89 kg/day
Concentration	5–7 g Ni/liter
Recovery rate	90%
Molecular weight	155 g/mole $NiSO_4$
Electrodialysis stacks (DU III Asahi Glass)	
Membrane spacing	2 mm
Membrane size	0.49 m × 0.98 m × 1 m
Effective area	70%
Number of cell pairs	50
Operating conditions	
Current density	10 mA/cm^2
Electric power consumption	1.70 kWh/kg of $NiSO_4$ recovered
Tangential flow rate	3.0 cm/sec
Pressure loss	0.1 bar
Coulomb efficiency	78.5%
Duration	7 years × 300 days × 21 hr
Coefficients	
Investment	18.5
Pumps	2.3
Motor-pump efficiency	0.5
Ion transfer	2
Unit prices	
Membranes	35,000 Yen/m^2
Electricity	5 Yen/kWh
Nickel	611 Yen/kg $NiSO_4$

Source: Data from Ref. 8.

Table 14B   Treatment of a Nickel Effluent by ED:
Economic Analysis, in Yen/kg of Nickel Sulfate Recovered

Cost of	Abbrev.	
Investment	CUAI	166.35
Pumping	CUFP	0.06
Ion transport	CUTI	17.00
Labor (3% of CUAI)	CUFL	2.38
Total cost	CUI	185.73
Price of nickel purchase		611.00
Net profit, in yen/kg $NiSO_4$ recovered		425.27

Source: Data from Ref. 8.

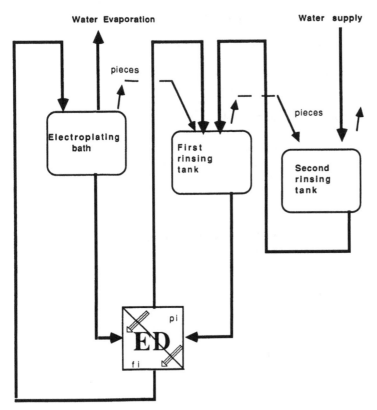

Figure 8 Concentration of copper from electroplating. Composition of electrodeposition bath and concentrate: copper ions, 65 g/liter; cyanide ions, 150 g/liter. Composition of first rinse bath and diluate: copper, < 1 g/liter. Water evaporation equals water supply, i.e., 50 liter/hr.

for evaporation from the electrolysis bath. The flux of water associated with the ion flux can be used to compensate for evaporation, as indicated in Figure 8.

An economic analysis for a French plant shows that savings of metal in this manner can pay back the investment within a year, as indicated in Table 15A and B.

Cyanide Baths

The treatment of effluents from cyanide baths is of great interest, since many surface treatments use this kind of solution (see, for example, Table 6). In a proposed system [39], wash tanks and electrodialyzers are connected in parallel in a multistage countercurrent system, as indicated in Figure 9. The effluent treated at each stage is circulated between the rinse tank and the electrodialysis stack as

Table 15A   Treatment of a Copper Effluent by ED:
Specifications

Effluent treated	
Copper content discharged	311 kg/month
Concentration	52 g Cu/liter
Recovery rate	94%
Molecular weight of Cu	63.5 g/mole
Electrodialysis stacks (SRTI)	
Membrane spacing	0.5 mm
Membrane size	0.5 m × 0.5 m × 1 m
Effective area	70%
Number of cell pairs	—
Operating conditions	
Current density	$-mA/cm^2$
Electric power consumption	1.50 kWh/kg of Cu recovered
Tangential flow rate	4.9 cm/sec
Pressure loss	0.1 bar
Coulomb efficiency	– %
Duration	1 year
Coefficients	
Investment	18.5
Pumps	2.3
Motor-pump efficiency	0.5
Ion transfer	2
Unit prices	
Membranes	1,000 francs/m^2
Electricity	0.15 francs/kWh
Copper	140 francs/kg Cu

Source: Data from Ref. 38.

Table 15B   Economic Analysis, in French francs/kg of Copper
Recovered

Cost of	Abbrev.	
Investment	CUAI	132.14
Pumping	CUFP	0.05
Ion transport	CUTI	2.80
Labor (1.5% of CUAI)	CUFL	1.98
Total cost	CUI	136.97
Price of copper purchased		140.00
Net profit, in French francs/kg of copper saved		4.03

Source: Data from Ref. 38.

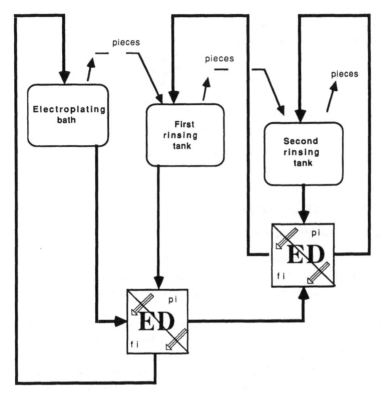

Figure 9 Elimination of cyanide from a galvanization effluent.

diluate. The concentrate is recycled to the previous bath, electrodeposition, or first rinse tank. Tests on this system in a Japanese facility are reported in Table 16. The concentration ratio of the concentrate solution to the dilute solution is approximately 70. As a result, the concentration of the effluent can be greatly reduced.

ED is often used to reduce the cyanide ion content in rinse water effluents from many basic galvanization baths, e.g., silver, copper, and zinc. But since the cyanide ion travels through membranes faster than metal ions, there is a discrepancy in the diluting solution between the cyanide ion content and the metal cation content. Moreover, in the diluate there is also an excess of hydroxyl ions since they pass through the anion membranes more slowly than the cyanide ions. As a result, some precipitation of metal occurs in the diluate, near the superficial zone of the anion exchange membranes. This failure can be easily eliminated by regulation of the cyanide content of the diluting solution to the correct value [40]. As a consequence, at least 96% of the metal discharged as ions into the wash solution is easily recovered.

Table 16 Treatment of a Cyanide Bath by ED

Composition of the galvanization bath treated	
CuCN	120 g/liter
NaCN	135 g/liter
Na_2CO_3	15 g/liter
NaOH	30 g/liter
Product solutions from the electrodialysis stack	
Diluted solution	0.520 g CN^-/liter
Concentrated solution	36 g CN^-/liter

APPLICATION TO OTHER INDUSTRIES

Glass Industry: Demineralization of Rinse Water

In glassware factories, much industrial water is generally consumed for cleaning of products. In some locations serious problems arise in obtaining enough high-quality rinse water at a low price, and in compliance with antipollution regulations.

One way to circumvent these problems is to recycle the major part of the wash water. However, this requires elimination of most of the dissolved and suspended matter before reuse. This can be solved by ED.

An example is the treatment of wash water in a Japanese glassware factory [49]. The desalination plant treats 480 tons/day of raw water, of pH 6.5–8, with an initial conductivity of 1200 $\mu S \cdot cm^{-1}$ and 750 ppm of TDS (total dissolved salt). The conductivity of the diluate from the staged desalination ED equipment is less than 200 $\mu S \cdot cm^{-1}$ and its TDS is lower than 80 ppm. So ED saves more than 83% of the water used.

The ED plant consists of a three-stage electrodialyzer with 400 membrane pairs in each stage, as indicated in Figure 10. One unit is used as the first stage, and the two remaining units are used as the second stage. Discharge from the manufacturing line is first treated by a sand filter which removes suspended matter. The filtrate is directly supplied to the first-stage diluate tank and recirculated to the first-stage tank. Overflow of the first-stage tank, which is about the same quantity as the supply of feed water, is fed to the second-stage diluate tank and recirculated in the second-stage units. Overflow from the second stage is returned to the manufacturing line. Part of the filtrate is supplied to the concentrating line, which works in the same way as the diluting stream. Finally, about 16% of the incoming raw water is discharged as concentrate.

With these considerations, a typical example of the operating costs during continuous operation is shown in Table 17A and B. It is noted that such a plant operates continuously without a full-time worker, and so labor costs can be omitted in the estimation. The cost of the installation is recovered in about 2 years.

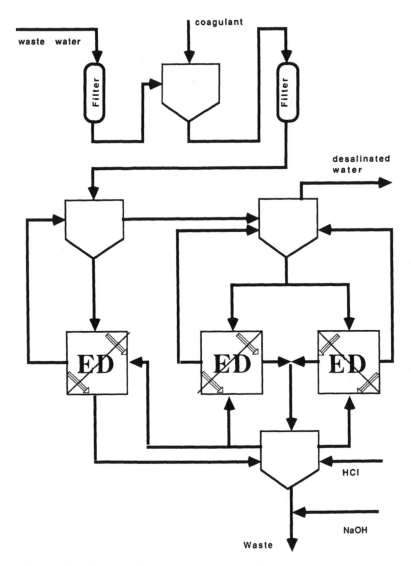

Figure 10 Two-stage desalination wastewater plant.

Table 17A Treatment of a Glassware Effluent by ED: Specifications

Effluent treated	
Water discharged	480 tons/day
Total dissolved salts (TDS)	750 ppm
Recovery rate	83%
Electrodialysis stacks (three DS-3 Asahi Glass)	
Number of cell pairs	400
Operating conditions	
Current density	$-$ mA/cm^2
Electric power consumption for ED	0.1625 kWh/ton H$_2$O recovered
Electric power consumption for pumps	0.875 kWh/ton H$_2$O recovered
Coulomb efficiency	$-$ %
Duration	2 years $\times$ 330 days $\times$ 24 hr
Chemicals	
35% HCl	1.675 kg/ton H$_2$O recovered
25% NaOH	0.0325 kg/ton H$_2$O recovered
Coagulant	0.0475 kg/ton H$_2$O recovered
Unit prices	
Membranes	$-$ Yen/m^2
Electricity	15 Yen/kWh
Water	200 Yen/m^3 H$_2$O
35% HCl	20 Yen/kg
25% NaOH	17 Yen/kg
Coagulant	100 Yen/kg

Source: Data from Ref. 48.

Table 17B Treatment of a Glassware Effluent by ED: Economic Balance, in Yen/m^3 Water

Cost of	Abbrev.		
Investment	CUAI		166.5
Operating cost of membrane replacement and maintenace	CUAR		11.0
Pumping	CUFP		12.2
Ion transport	CUTI		2.3
Chemicals			
35% HCl		3.1	
25% NaOH		0.5	
Coagulant		4.4	
Total	CUVM		8.0
Total operating costs			33.5
Price of water purchased			200.00
Net profit, in yen/ton water saved			166.5

Source: Data from Ref. 48.

Chemical Industries: Recovery of Catalyst

In the chemical industries, electrodialysis is generally used for the concentration of electrolytes in aqueous effluents. But, as indicated in the section "Integrated with the Main Operation," the best way to include the ED operation is within the process itself and not at the end of the flowsheet. Some confidential applications are currently running in several chemical plants.

A typical example is a proposal for the recovery of a catalyst used in an organic reaction [41]. Hexanediol is an alcohol used in the production of polymers. It can be obtained from propargyl alcohol according to the following route: hexadiine (2,4)-diol(1,6) is produced from propargyl alcohol by oxidative addition employing mixed inorganic salts, such as copper and ammonium chloride, as catalysts. The diol is then converted to hexanediol by hydrogenation according to:

$$2HC{\equiv}C{-}CH_2{-}OH + \tfrac{1}{2}O_2 \rightarrow [OH{-}CH_2{-}C{\equiv}C{-}]_2 + H_2O \qquad (48)$$

$$[OH{-}CH_2{-}C{\equiv}C{-}]_2 + 3H_2 \rightarrow HO{-}(CH_2)_6{-}OH \qquad (49)$$

However, the relatively low solubility of the diine diol in water leads to its precipitation in an aqueous medium, which presents some difficulties in its recovery. Furthermore, the removal of the salts is a problem since extraction with water amounts to product loss. Desalination by means of ion exchangers is not economical as the salt content in the solution is too high for such a method.

The method presented in Figure 11 involves conversion of the propargyl alcohol to diine diol in the presence of n-butanol. The diol product is found in the organic phase, in addition to a certain amount of salts, amounting to 0.1–2 wt %. Most of the salts remain in the aqueous phase and can be employed in the next batch.

Removal of the salts inevitably present along with water in the aqueous phase can be effectively carried out by electrodialysis as indicated in Figure 11. In this way, the salts are picked up by the reaction mixture itself and are recycled back to the reactor for the next batch operation, thus avoiding salt discharge into the environment as well as use of fresh salts for the next operation. In other words, electrodialysis leads to recovery and recycling of salts, and allows "zero discharge."

Experiments with reaction mixtures of diinediol, n-butanol, and dissolved salts and water were carried out in a stack with 50 cells of 0.4-mm thickness. The critical voltage was found to be on the order of 4 V per cell pair. Desalting by electrodialysis was conducted at 2 and 3 V per cell pair and 0.05–1 A. No fluxes of water or organic compounds were found between the diluting and the concentrating streams. It was found to be advantageous to carry out most of the salt removal employing only half the available amount of concentrate (i.e., a fresh mixture of propargyl alcohol and n-butanol), and following up the step with a purification employing the other half. The electrical energy necessary for the process in on the order of 3 kWh/ton of hexanediol. Effectively, the recovered salts were reusable as catalysts.

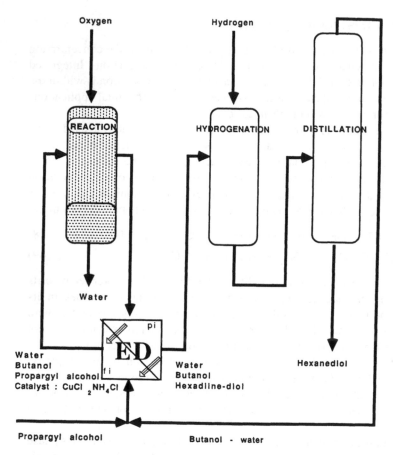

Figure 11 Recovery of the catalyst in the production of hexanediol.

Food Industry: Treatment of Wheys and Other Milk Byproducts

The mass production of cheese has made the discharge of whey a major pollution problem for rivers. The pollution of water by milk products is the result of fermentation in water of the organic matter contained in the effluents. This fermentation, which is aerobic, develops by consuming the dissolved oxygen normally present in water, resulting in suffocation of fish and promotion of particular flora to the detriment of the natural aquatic flora. For example, in France the fauna has totally disappeared from some rivers in Normandy.

Fortunately, all the byproducts of milk, such as buttermilk or wheys, have high protein contents, traditionally reserved for feeding pigs. For example, a typical French cheese-making or casein factory discharges about 200 tons of whey per day, containing 12 tons of dry matter of high nutrient value. American dairies discharge nearly 10 million tons of whey per year containing about 600,000 tons of dry matter. The corresponding water pollution equals that caused by a population of 2 million people. However, the protein content of this quantity of dry matter would be sufficient for the dietary needs of 8 million people. Thus, since whey has a high food value, the milk industry started 20 years ago to use it in animal feed and subsequently in human food. As a consequence, milk byproducts are no longer a pollution factor but are a good positive example of treatment of waste by ED.

Wheys can be separated into two main groups:

Sweet wheys, from cheese making, with an acidity less than 16° Dornic; 1° Dornic corresponding to 100 mg of lactic acid per liter

Acid wheys, with an acidity from 20–75° Dornic, from cheese whey plants, containing acid, or from casein production, with hydrochloric or sulfuric acid

The average composition of whey is indicated in Table 18A. It may be seen in this table that wheys have a fairly high salt content. An excess of these salts can cause fatal diarrhea in calves and in babies. Moreover, their acid content precludes any use as a foodstuff for cattle. Furthermore, when their acidity exceeds 40° Dornic, they cannot be reduced to powder by concentration (e.g., by atomization), and the product is highly hygroscopic.

So it is clear that the objective of ED should be to deacidify and to demineralize these two kinds of wheys and to bring their content of organics and minerals, as well as their acidity, down to values suitable for feedings animals or people.

Processing conditions are such that it is possible to treat raw whey (Table 18B), with a total dry matter content of 60 g/liter, or preconcentrated whey, with a total dry matter up to 180 g/liter. Generally, the process runs in a batch mode.

With the above considerations, a typical example of costs is shown in Table 18C for a preconcentrated whey [53]. This estimate is made for an amortization period of 5 years, in common use in most of the hundreds of plants all over the world, but the turnover of investments is generally shorter.

Food Industries: Liquid Waste from a Distillery

Wastewater from food industries generally contains large quantities of organic matter. They are treated with great difficulty by conventional methods, based on phase equilibria, such as centrifugation, crystallization, decantation, evaporation, etc., which are always very expensive. Also, the simultaneous presence of salts and suspended matter lowers the efficiency of a single process.

So it is necessary to develop integrated membrane processes for the separation of pollutants in such a way as to reuse valuable byproducts and to fall within allowable limits.

Table 18A Treatment of a Preconcentrated Whey by
ED: Normal Composition of Wheys, in g/liter

Component	Cheese whey	Casein whey
Water	930	930
Lactose	49	51
Proteins	70–10	70–10
Mineral salts	6	8
Soluble nitrogen matter	1.5	1.5
Fatty matters	1–2	1–2
Lactic acid	2	0
Total dry matter	63	70

Source: Data from Ref. 53.

Table 18B Treatment of a Preconcentrated Whey by ED: Specifications

Effluent treated
    Whey discharged   30 tons/day
    Total dry matter   18%
    Demineralization   90%
    Deacidification   50%
    Recovery rate   83% (4500 kg powder/day)
Electrodialysis stacks (SPF–Corning)
    Capacity of plant   90 tons/day of 6% dry matter whey
    Number of cell pairs   700
    Size   250 × 250 mm × mm
    Effective area   260 m²
Operating conditions
    Current density   −mA/cm²
    Electric power consumption for ED   0.112 kWh/kg of powder recovered
    Electric power consumption for pumps   0.063 kWh/kg of powder recovered
    Coulomb efficiency   − %
    Duration   5 years × 250 days × 22 hr
Chemicals
    35% HCl   24.5 kg/ton of powder recovered
    Washing agent   1.6 kg/ton of powder recovered
Unit prices
    Membranes   − FF/m²
    Electricity   0.35 FF/kWh
    Water   − FF/m³ H₂O
    35% HCl   1 FF/kg
    Washing agent   18 FF/kg

Source: Data from Ref. 53.

Table 18C Treatment of a Preconcentrated Whey by ED: Economic Analysis, in French francs/ton of Powder

Cost of	Abbrev.		
Investment	CUAI		18.66
Operating cost of membrane replacement and maintenace	CUAR		5.09
Pumping	CUFP		22.40
Ion transport	CUTI		39.20
Chemicals			
35% HCl		24.50	
Washing agent		28.80	
Total	CUVM		52.30
Labor	CUFL		70.00
Total operating costs			207.00
Price of whey, demineralized and concentrated by atomization			3750.00
Net profit, in French francs/ton powder saved			High

Source: Data from Ref. 53.

A typical example is the experimental work carried out with wastewater from an Italian alcohol industry. This water contains 1 wt % of proteins, 3.6 wt % of dissolved salts, and 8.6 wt % of suspended matter [42]. According to the process shown in Figure 12, the solution is first filtered through a hollow-fiber microfilter to remove suspended matter. Recovery of up to 95% is obtained and 5% of the feed is then rejected and sent to a decanter for further thickening.

The filtrate from the microfilter is ultrafiltered with a suitable membrane in order to split it into a liquid permeate of high salt content and a concentrated solution with a high organic content. Both are further concentrated.

The permeate is concentrated by low-pressure reverse osmosis and electrodialysis for recovery of salts as solids. The ultrafiltered concentrate is further concentrated by high-pressure reverse osmosis and the proteins obtained are used as an additive in animal feed. All of the clean water collected is reused in the preceding steps of the process.

With such an integrated process, it is clear that the dilute waste from the distillation steps is divided into two concentrate streams, one rich in protein, the second formed by solid salts. The remaining water can be used anywhere pure water is needed.

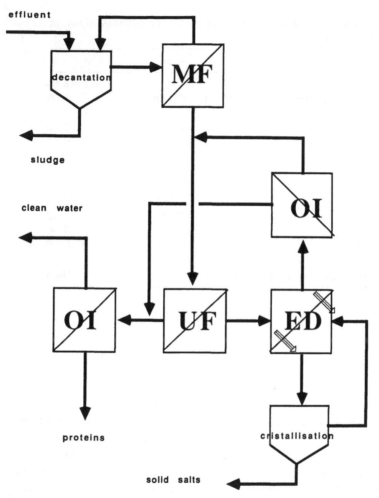

Figure 12   Wastewater from a distillery treated with associated membrane process.

MEMBRANE ELECTROHYDROLYSIS

Bipolar Membranes

Bipolar membranes are able to separate water into hydrogen and hydroxyl ions. When used in conjunction with conventional cation and/or anion exchange membranes in a cell stack, they provide a low-energy process for generating the acid and the base from the salt.

Figure 13 shows a bipolar membrane in a single composite film consisting of

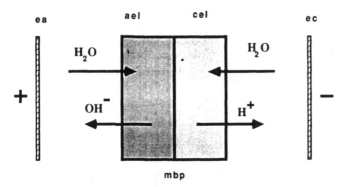

Figure 13 A schematic bipolar membrane. mbp, bipolar membrane; cel, cation exchange layer; ec, cathode; ael, anion exchange layer; ea, anode.

two layers of oppositely charged ion exchange materials. One layer is permeable to anions and the other to cations. In an aqueous solution, when a direct current is passed across the bipolar membrane with its cation-selective side toward the cathode, any salt which diffuses into the region between the two layers is removed. Moreover, in this case, the two layers of the membrane are located in such a manner that under the action of the direct electric field the transport of anions and cations toward the inside of the membrane is hindered.

As a consequence, even at relatively low current densities, the concentration of the salt ions inside the membrane falls to such a low level that the majority of ions available to carry the current are the H^+ and the OH^- ions, which are always available at around 10^{-7} mole/liter from the dissociation of neutral water. Moreover, the ion exchange layers are highly water-swollen, and since water is uncharged, it can diffuse rapidly through the membrane from either side to the inside and so replace the water being dissociated.

In order for the membrane to be useful in acid–base generation or purification, it must be used in conjunction with monopolar membranes in a stack of the same kind as an ED stack. Figure 14 shows a three-compartment arrangement for converting NaCl to HCl and NaOH. In practice, a large number of such repeating elementary units is placed between a single set of electrodes, as in an ED stack. But because there are many more operating elementary units than electrodes, the effect of electrodes on the overall stack characteristics is minimal. The inherent thermodynamic advantage of a membrane water splitter results from its ability to generate NaOH and HCl with a power consumption of about 1.23 kWh/kg of water dissociated.

Moreover, a membrane electrohydrolysis stack is generally used in conjunction with an electrodialysis stack in order to provide the right concentration in a loop.

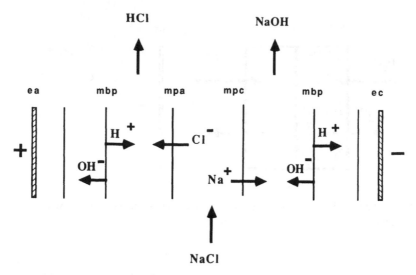

Figure 14 Electrohydrolysis of sodium chloride with bipolar membranes in a three-compartment water dissociation device. mpa, anion exchange membrane; mpc, cation exchange membrane; mbp, bipolar membrane.

Recovery of Steel Pickle Liquors

Mixtures of dilute hydrofluoric acid (HF) and of nitric acid (HNO_3) are used in the pickling of stainless steels to remove oxide scale and chromium-depleted layers formed during the annealing step in steel- making processes.

In a process operating in a North American steel factory, the concentrations of pickle acids used in a pickle line are initially 2–3.5 wt % HF and 6–10 wt % HNO_3, depending on the grade of steel undergoing treatment [43,47]. The initial step of the process is the neutralization of the spent pickle acid with potassium hydroxide. Neutralization of the free hydrofluoric and nitric acids in the effluent forms the associated potassium salts indicated in the reactions:

$$HF + KOH \rightarrow KF + H_2O \tag{50}$$

$$HNO_3 + KOH \rightarrow KNO_3 + H_2O \tag{51}$$

Studies have shown that the salts present in the spent pickle effluent are in the form of complexed metal fluorides rather than as free acid. The principal metals produced from stainless steel pickling are iron chromium, and nickel. Their fluorides are converted to metal hydroxides and they are recovered in a relatively dry filter cake suitable for recycling to the steel-processing facility.

$$FeF_3 + 3KOH \rightarrow 3KF + Fe(OH)_3 \tag{52}$$

$$CrF_3 + 3KOH \rightarrow 3KF + Cr(OH)_3 \qquad (53)$$

$$NiF_2 + 2KOH \rightarrow 2KF + Ni(OH)_2 \qquad (54)$$

After filtration, the clean potassium salts are processed in the electro-hydrolysis step. For this the ion exchange membranes are arranged in a three-compartment cell stack configuration, as indicated in Figure 14.

Electric current applied across the stack results in the conversion of potassium salts to the mixed HF and HNO_3 acid and KOH base:

$$KF + H_2O \rightarrow HF + KOH \qquad (55)$$

$$KNO_3 + H_2O \rightarrow HNO_3 + KOH \qquad (56)$$

The three circulation loops operate in a batch mode in order to achieve the highest operating efficiency.

As indicated in Figure 15, both free and complexed fluoride and nitrate components are recycled to the pickle line as mixed acids at the required volume and concentration. The potassium component is internally recycled to the neutralization step as potassium hydroxide after concentration, and the diluate is used to wash the filter cake after demineralization by ED. The concentrate is recycled at the inlet of the electrohydrolysis stack to attain the right concentration in the effluent streams, e.g., 1.5 M for KOH. In this way the reduction of the quantity of hydrofluoric acid purchased is about 60%, that of nitric acid is about 40%, and spent acid disposal is reduced by 70%.

With these considerations, some operating conditions are given in Table 19A and B. In this table, costs are indicated for a plant with a disposal of 6057 m^3 of spent acid per year, in the case of no treatment, of a classical purification loop and of an electromembrane loop, using electrohydrolysis and electrodialysis.

It can be easily seen that any treatment is more expensive than the lack of treatment, but now environmental pressures do not allow such a route. The unit costs of the two treatments are not very different, but it is obvious that the sum of the operating and chemical costs is lower in the case of the electromembrane process than in the case of a classical treatment. It is also lower than in the case of the absence of treatment. So the amount saved in operating costs allows a payback within less than 2 years.

Recovery of Glass-Etching Solution

More and more industries use silicon and its derivatives for various purposes. During the manufacture of parts or objects, fluorinated compounds are often necessary to make specific transformations. Generally, it is a surface treatment. As a result of the different operations, aqueous wastes contain variable quantities of fluorosilicic acid issued from the plant.

For example, in a glassworks, surfacing, frosting, and etching are carried out

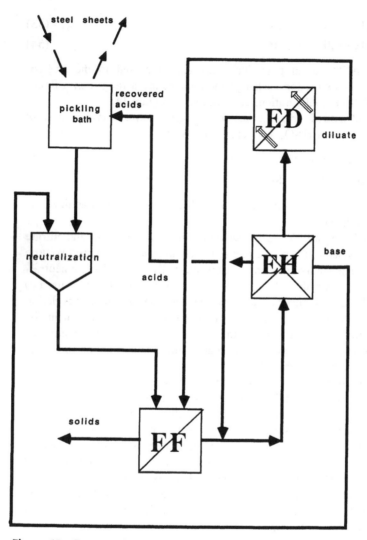

Figure 15 Regeneration, concentration, and reuse of nitric acid–hydrofluoric acid mixture used in steel pickling.

in a bath containing ammonium bifluoride. By reaction with vitreous silica, hexafluorosilicic acid is formed:

$$SiO_2 + 6HF \rightarrow H_2SiF_6 + 2H_2O \qquad (57)$$

In a pilot plant of a French glassware factory, using 100 metric tons (100% HF) per year of fluorinated compound, the effluent from a methodic rinse, i.e., from a

Table 19A Treatment of a Pickle Liquor by EH and ED: Specifications

Effluent treated	
Spent acid disposal	6057 m³/year
Total concentration of HF	9.38 wt %
Total concentration of HNO₃	15.00 wt %
Total concentration of iron	3.60 wt %
Liquor density	1.1
Electrodialyzer stack (Aquatech)	
Electrohydrolysis stack (Aquatech)	
Unit prices	
Hydrofluorohydric acid 100%	1,500 US$/ton
Nitric acid 100%	250 US$/ton
Electricity	0.045 US$/kWh
Tap water	− US$/m³ of H₂O
Spent acid disposal	132 US$/m³
Sludge disposal	500 US$/ton
Cost recovery period	3 years

Source: Data from Ref. 43.

Table 19B Treatment of a Pickle Liquor by EH and ED: Economic Balance, US$/m³ Spent Acid

Cost of	No treatment	Purification	Electromembrane
Investment	0.00	61.91	162.35
Membrane maintenance	0.00	0.00	49.53
Energy	0.00	0.49	24.76
Equipment maintenance	0.00	2.48	16.50
Labor	0.00	0.49	33.02
Total for operating costs	0.00	3.46	123.81
HF purchase	173.35	173.35	74.29
HNO₃ purchase	55.72	55.72	33.02
Spent acid disposal	132.08	132.08	41.27
Sludge disposal	20.63	20.63	20.63
Increase of acid consumption (20%)	0.00	26.41	0.00
Chemical cost total	381.78	408.19	169.21
Operating cost total	381.78	411.65	293.02
Total cost	381.78	473.56	455.37

Source: Data from Ref. 43.

countercurrent washing machine, is first neutralized with an alkaline base or ammonium hydroxide, which precipitates silicon as an insoluble hydroxide, and which retains fluorine as a soluble ion according to the following reaction [45]:

$$H_2SiF_6 + 6NH_4OH \rightarrow Si(OH)_4 + 6NH_4F + 2H_2O \tag{58}$$

The silica precipitate is separated from the clear solution by filtration. The solution containing the fluorinated salts is then concentrated by electrodialysis, which yields a diluate recycled at some point in the rinsing operation, and a concentrate used in the stack fitted with bipolar membranes, where electrohydrolysis occurs. Electrohydrolysis regenerates the starting base and the hydrofluoric acid.

$$NH_4F + H_2O \rightarrow NH_4OH + HF \tag{59}$$

The solution containing the base is recycled to the neutralization tank and the solution containing the hydrofluoric acid is sent to a second electrodialyzer. The second electrodialysis operation concentrates the acidic solution to a level essential for its reuse in the glass-corroding bath.

As indicated in Figure 16, the losses of process water, hydrofluoric acid, and neutralizing base are eliminated. So nothing is discharged into the environment. The savings in hydrofluoric acid pays back the plant in less than 10 years, in

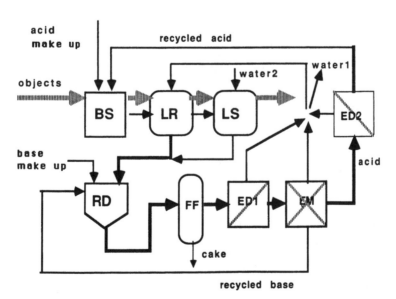

Figure 16 Process for the recovery of a fluorinated solution. Normal process: LR, LS = washing bath. Membrane process: ED1, ED2 = electrodialysis; BS = etching bath; RD = chemical neutralization; FF = frontal filtration; EM = electrohydrolysis.

Table 20A Treatment of the Etching Solution by EH and ED or by Classical Precipitation: Specifications

Effluent treated	
Spent acid disposal	32,000 m³/year
Total concentration of fluor	2.7 wt %
Total annual sludge discharged	1265 tons/year
Electrodialyzer stacks (two)	
Electrohydrolysis stack (Aquatech)	
Unit prices	
Hydrofluorohydric acid 100%	8.70 French francs/kg
Electricity	0.36 French francs/kWh
Tap water	5.00 French francs/m³ H_2O
Sludge transport	258 French francs/ton
Sludge disposal	225 French francs/ton
Cost recovery period	10 years

Source: Data from Ref. 44.

comparison to a classical precipitation with calcium lime, as indicated in Table 20A and B.

In this case, it can be seen that the membrane process is less expensive than the lime process or no treatment. The unit costs of the two treatments are very different, but it is obvious that the sum of the operating and chemical costs is still lower in the case of the electromembrane process than in the case of a classical

Table 20B Treatment of the Etching Solution by EH and ED or by Classical Precipitation: Economic Analysis in French francs/m³ of Effluent

Cost of	No treatment	Purification	Electromembrane
Investment	0.00	6.50	25.60
Membrane maintenance	0.00	0.00	2.60
Energy	0.00	0.46	0.46
Equipment maintenance	0.00	0.65	0.56
Labor	0.00	3.25	1.28
Total for operating costs	0.00	10.86	30.50
HF purchase	23.50	23.50	0.00
Other chemicals purchased	0.00	9.30	0.00
Sludge transport	0.00	10.20	0.00
Sludge disposal	0.00	8.90	0.00
Taxes for pollution	23.50	3.00	0.00
Chemical cost total	47.00	54.90	0.00
Operating cost total	47.00	59.26	4.90
Total cost	47.00	65.76	30.50

Source: Data from Ref. 44.

treatment, but the investments are higher. As a consequence, the amount saved in operating costs pays back the installation of the membrane process within less than 10 years.

ELECTRO-ELECTRODIALYSIS

The Principle

In some cases, the cells in an ED stack are restricted to the two-electrode chambers. But the ion exchange membrane between the two electrodes remains, as in a membrane electrolysis cell. According to the objective, the membrane fitted in the cell is either an anion exchanger or a cation exchanger. As in an ED stack, under a direct current potential difference applied to the electrodes, coions travel through the membrane as indicated in Figure 17.

At the same time, electrode reactions occur. For example, if the solution treated is an aqueous mixture, the water is dissociated. But as electrons are exchanged between the electrode and the components of the solution, only one ion can be generated (proton or hydroxyl ion), with a gas (oxygen at the anode or hydrogen at the cathode), according to the value of the pH of the solution. Table 21 indicates the different electrode reactions, when only the dissociation of water occurs.

However, as the aqueous solutions treated by electro-electrodialysis (EED) contain a salt resulting from the reaction of a base with an acid, some electro-chemical oxidation of the anion frequently occurs. So new components can react.

For example, in the case of electrolysis of a natural brine, chloride ions may be oxidized at the anode to produce hypochlorite according to the following reaction:

$$Cl^- + 2OH^- \rightarrow H_2O + OCl^- + 2e^- \tag{60}$$

In an acidic solution, chlorine gas will be released:

$$2Cl^- \rightarrow Cl_2 + 2e^- \tag{61}$$

A portion of the chlorine oxidized at the anode will be transported from the cell in the anode rinse stream, or anolythe, and a portion may escape as chlorine gas. Chlorine remaining in the anolythe will eventually form an equilibrium mixture according to:

$$Cl_2 + H_2O \rightleftarrows HOCl + HCl \tag{62}$$

As a consequence, the membranes used, anion or cation exchange, will be more resistant to oxidation and to oxidizing substances than normal ED membranes.

Recovery of Chromium by EED

It is advantageous to recover chromium from waste effluents of electrodeposition baths, even though chromic acid is cheap ($\sim$ 25 French francs/kg), since large

chromic acid to chroming bath

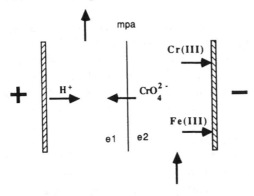

Figure 17   Principle of an electro-electrodialysis (EED) cell.

quantities are used for corrosion protection and for decorative purposes. Moreover, chromium ions, and especially Cr(VI) ions, are very toxic.

In a simple cell, an anion exchange membrane is located between two electrodes, anode and cathode, as indicated in Figure 17. This membrane allows negatively charged chromate ions to be separated from cations, such as Fe(III) and Cr(III).

The aqueous solution containing acidic chromium ions is fed into the catholythe compartment. The chromate ions, which can cross the anion exchange membrane as a coion, form chromic acid by combining with protons generated on the anode. Combination of dialysis and electrolysis allows very concentrated acid up, to 280 g/liter, to be obtained. It is then suitable for recycling to the chroming bath. However, the energy expenditure is always high, on the order of 10–20 kWh/ kg of CrO_3 recovered.

The anion exchange membrane works under very severe conditions, and thus it may be frequently replaced. In order to reduce the cost of membrane maintenance, some special fittings have been designed especially for the chroming process.

Table 21   Electrode Reactions of Water

Cell	Anode	Cathode
Acidic	$2H_2O \rightarrow O_2 + 4H^+ + 4e^-$	$2H^+ + 2E^- \rightarrow H_2$
Basic	$4OH^- \rightarrow O_2 + 2H_2O + 4e^-$	$2H_2O + 2e^- \rightarrow H_2 + 2OH^-$

One typical application is the concentration and the recycling of chromic acid in hard chromium plating lines, but applications also exist for decorative chroming [50].

A chromic bath is normally loaded with chromic acid at concentrations of 250–300 g CrO_3/liter. During the plating operation, this bath loses acid by electrolytic deposition, adherence to chromed parts or articles, and misting. But since generally the temperature is about 35°C, evaporation is not a serious problem, and the concentration of the bath may be maintained constant by addition of fresh chromic acid. Contaminants such as Fe(III), Cu(II), Al(III), and Cr(III), arising from the metallic parts treated, are dragged out in the rinse tanks.

To avoid the increase in concentration of chromium in the first rinse tank or static rinse tank, which can reach 150 g CrO_3/liter in a traditional batch process, an EED cell is fitted between this tank and the chroming bath, as indicated in Figure 18. More than 90% of the chromic acid dragged out by pieces is saved and returned for use at the right concentration. As a consequence, in the static bath, chromium is more dilute and its concentration is easily lowered to 20 g CrO_3/liter.

Moreover, after the static rinse, there is a countercurrent rinse line, and the effluent is generally treated by ion exchange resins. The anion exchange resin is regenerated with caustic soda, and chromium is recovered as sodium chromate. EED then treats the chromium lost in the anionic demineralizer resin backwash and permits return of a purified soda solution for the next anionic ion exchanger regeneration.

With these considerations, some operating conditions are given in Table 22A and B. In this table, a comparison is made only on the operating costs between a traditional plant with and without EED loops for reuse and concentration of wasted chromic effluents. It appears that the difference in net profit between both processes, with and without EED, is on the order of 20 French francs per kg of CrO_3 recovered. Generally, this profit is enough to pay back the new investment within a few years.

DIALYSIS

Ordinary and Donnan Dialysis

In a broad sense, dialysis refers to a process separating one chemical species from another species in a liquid solution through a semipermeable membrane. The driving force is the difference of the chemical potential across the membrane.

The selectivity of the separation between the two chemical species in a membrane depends on the relative interactions between the membrane and the species.

In the case where steric effects are the principal factors, the selectivity of the membrane is related to the ratio of the molecular sizes of the chemical species. In this sense, the process has been applied to separate small crystallites from large

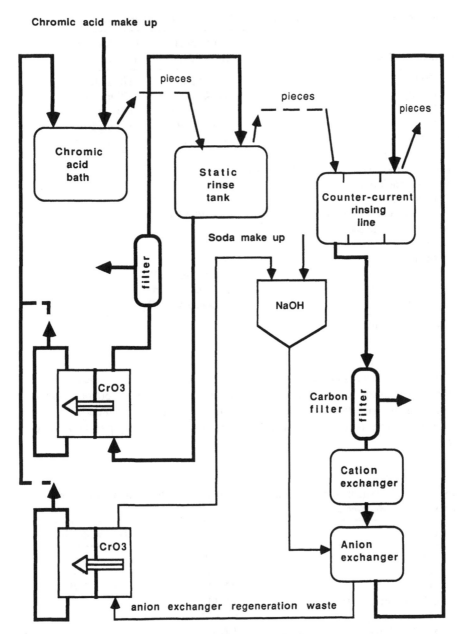

Figure 18 Concentration and reuse of chromic acid by EED.

Table 22A Concentration of Chroming Efflulents by EED: Specifications

Solutions involved	Operating chromium bath	EED Anolythe recycled loop
CrO$_4$	250 g/liter	270 g/liter
Cr(III)	1,075 mg/liter	130 mg/liter
FE(III)	250 g/liter	100 mg/liter
Cu(II)	850 g/liter	310 mg/liter
Ni(II)	2,500 g/liter	875 mg/liter
Recovery rate	—	90% of CrO$_4$
Electro-electrodialysis stack (EIF Ecologie)		
Operating conditions		
Current density	—	
Electric power consumption	15 kWh/kg CrO$_4$ saved	
Coulomb efficiency	—	
Unit prices		
Membranes	—	
Electricity	0.35 francs/kWh	
Chromic acid	13 francs/kg CrO$_4$	
Wasting	13 francs/kg CrO$_4$	

Source: Data from Ref. 50.

colloids in solution. However, the permeation flux of the smaller species with the driving force of the concentration gradient alone is often found to be too small to be of economic significance for low-cost components. As a consequence, the use of ordinary dialysis, or molecular dialysis, is restricted to very particular mixtures.

Table 22B Concentration of Chroming Efflulents by EED: Economics, in French francs/kg CrO$_4$ Recovered

Cost of	Abbrev.	Without EED	With EED
Investment	CUAI	—	—
Pumping	CUFP	—	—
Ion transport	CUTI	—	5.25
Labor (% of CUAI)	CUFL	—	—
Chromic acid purchase	CUVM	13.00	—
Spent acid disposal	CUVE	13.00	
Resin exchange	CUFT	—	—
Total operating costs		26.00	5.25
Net profit, in French francs/kg CrO$_4$ saved			20.75

Source: Data from Ref. 50.

If the main interaction between chemical species and membrane materials is due to ionic effects, Donnan equilibrium is maintained at the membrane-solution interface. This effect prevails when electrolyte solutes are dialyzed through neutral membranes or ion exchange membranes. It is common to limit the term of Donnan dialysis to the dialysis of electrolyte solutes with ion exchange membranes.

Dialysis of Acids

In industry, diffusion dialysis (or Donnan dialysis) is mainly used to recover acids from effluents from hydrometallurgical or galvanization plants, as indicated in Table 23. It is also known as acid dialysis.

In a dialysis stack used for acid recovery, only anion exchange membranes, mainly permeable to anions, are fitted in a filter press assembly, quite similar to the electrodialysis stack, described in the section "Principle of Electrodialysis." But unlike an electrodialyzer, all the membranes are now of the same kind (i.e., in anion exchange membranes), and there are no electrodes at the ends. The anion exchange membranes used are not very different from those used in ED, but they are generally less selective to anions.

The driving force between the diluting compartment and the acid recovering compartment is the gradient of the chemical potentials of anions which are well-known functions of the concentrations [26]. Generally the two kinds of compartments so formed are fed in parallel.

According to Figure 19, under a difference in concentration of anions between the diluting compartment d and the recovering compartment b anions of the effluent leave compartment d7 by passing through membranes a6 and a7 and reach compartments b6 and b8. But as concentrations of anions in these two compartments, b6 and b8, are lower than in the next compartments d5 and d9, they cannot leave them.

Moreover, this migration of anions creates an imbalance in the electroneutrality of the solution in compartment d7, as well as in compartments b6 and b8. So

Table 23   Principal Applications of Diffusion Dialysis

Waste acid from pickling of steel, stainless sheets	H_2SO_4, HCl, HNO_3, HF
Refining of waste acid from batteries	H_2SO_4
Recovery of metal in refining process	H_2SO_4, HCl, H_3PO_4
Treatment of waste acid from aluminite processing	H_2SO_4, HNO_3
Waste acid from etching of Al, Ti, glass	HCl
Waste acid from surface treatment in plating line	HCl, H_2SO_4, HNO_3, HF
Treatment of waste acid in deacidification of refining in organic synthesis processes	HCl, H_2SO_4

Source: Data from Ref. 46.

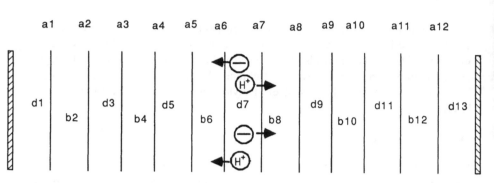

Figure 19 Principle of acid transfer by dialysis. a, anion exchange membranes; d, diluting compartment; b, recovering compartment.

cations also tend to leave compartment d7. But according to the selectivity of the anion exchange membranes used, usually cations cannot pass through membranes a6 and a7. However, protons (H$^+$) travel easily through the anion exchange membranes (a6 and a7) and enter compartments b6 and b7. As a consequence, both ions forming the acid, protons and anions, leave compartments d and enter compartments b. The anion exchange membranes are used so that these transfers of protons are facilitated.

Transfer of material occurs only when the concentration of acid in compartments b is lower than in compartments d. So this transfer of matter diminishes progressively as both concentrations, in b and d, equalize. In common practice, however, there always exists some difference between these two concentrations, as the two solutions flow continuously in parallel through compartments b and d.

Basic Equations

Donnan dialysis is essentially a continuous ion exchange process. As there is a difference in concentration in an ionic species across the membrane between the two streams, an electric field is created. Under suitable conditions, the potential, in turn, greatly enhances the permeation rate of another ionic species. Even "pumping" of the other ionic species is possible from a less concentrated solution into a more concentrated solution. This is the basis of liquid membranes which are not examined here.

Then, the initial set of equations is the same as indicated in the section "Basic Relations" for electrodialysis. But as there is no external electric field imposed, the current density j is zero. As a consequence, the flux of the counterion i passing normally through the membrane permeable to it is simply:

$$J_i = -D_i C_i K_m \, \text{grad} \, (\ln a_i) \tag{63}$$

which is only a diffusion flux. The coefficient K_m takes into account the properties of the ion i within the membrane.

Generally, in acid dialysis, even if the solutions are not too concentrated, it is not possible to use the concentrations C_i instead of the activities a_i; as in acidic solutions, the activity coefficients are generally very different from unity [26].

Moreover, in a dialysis stack, as in the ED stack, it is necessary to take into account the resistance to transport occurring in the two concentration polarization layers, R_{id} and R_{ib}.

However, for the sake of simplification, if we use concentrations instead of activities, at steady state, the flux of counterion i is given by:

$$J_i = \frac{1}{R_{Ti}} (C_{id} - C_{ib}) \tag{64}$$

where the total resistance R_{Ti} is a function of the resistances of the concentration polarization layers R_{id} and R_i and $D_i K_m$. It is expressed as:

$$R_{Ti} = R_{id} + \frac{1}{D_i K_m} + R_{ib} \tag{65}$$

Finally, it appears that the possibility of using dialysis to recover an acid through an ion exchange membrane imposes the reduction of the overall resistance R_{Ti}. This must be done by increasing the term $D_i K_m$ resulting from a combination of the properties of the membrane, and by diminishing the resistance of each concentration polarization layer resulting from a good choice of the operating conditions and of the design of each circulation compartment.

However, Eq. 63 is valid only over a very narrow zone of the dialysis membrane, where concentrations remain constant. As Donnan dialysis is generally used in the counterflow mode, as indicated in Figure 20, the variations in concentration of solutions through the dialyzer must be taken into account. So the driving force over the stack is the mean logarithmic difference in concentrations. This is expressed by Eq. 66 where C_{df}, D_{do}, and C_b are respectively the concentrations of waste acid feed, of purified waste, and of recovered acid in the case where pure water is used as recovering fluid.

$$\Delta C = \frac{C_{df} - C_b - C_{do}}{\ln \left[(C_{df} - C_b)/C_{do} \right]} \tag{66}$$

The ionic flux is usually expressed by Eq. 67, where U is known as the dialysis coefficient.

$$J_i = U \cdot \Delta C \tag{67}$$

For the dialysis membranes currently available, the dialysis coefficient U for useful acids is between 1 and 10 mol of acid per hour, square meter and (mol per

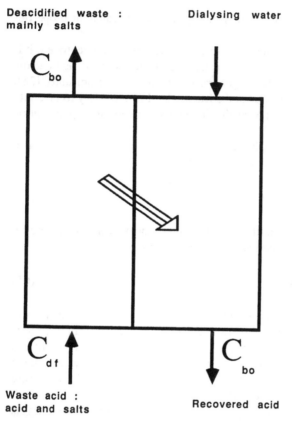

Figure 20   Principle of countercurrent dialysis.

liter), and it is 10–100 times greater for the salts encountered in practice with these acids.

Dialysis of Effluents from a Nickel Galvanization Bath

Because of the usefulness and cost of nickel, it is profitable to recover. This can be done by ED, as indicated in the section "Basics of the Balances." In a surface treatment process, if the quantity of nickel discharged from the galvanization tank is small, i.e., when the nickel concentration in the effluent obtained after washing is low, an ion exchange resin column can be used to treat the effluent solution, as indicated in Figure 21. Once the nickel is bound to the resin, the purified water can be reused as wash water. The resin-bound nickel is eliminated by an acid, regenerating the resin. The nickel contained in the eluted acid solution is then

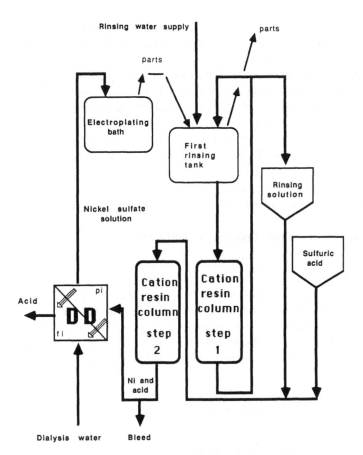

Figure 21 Dialysis of sulfuric acid from a nickel sulfate solution.

recovered by dialysis in order to eliminate coexisting excess acid. The nickel solution is then returned to the electroplating bath, as indicated in Figure 21.

With such a system in a Japanese facility, more than 99% of the free sulfuric acid is eliminated by using an excess of water relative to the quantity of effluent treated. This excess water allows the existence of a sufficient difference in concentrations between the two compartments, diluting d and recovering b. But this excess water also reduces the thermal effect due to the introduction of a strong acid in water [26].

The effluent obtained by elution of the resin with sulfuric acid is divided into two parts. The final portion, with a low nickel content, is stored as supply water for the next resin regeneration. The initial portion, with a high nickel content, is treated by diffusion dialysis.

The nickel sulfate solution obtained by diffusion dialysis is slightly acidic and contains 20–30 g of Ni per liter. It is then recycled to the electrodeposition bath after its composition has been adjusted.

This nickel sulfate solution is recovered in a form completely free of organic material. But cations coexisting with nickel are not eliminated. As a consequence, deionized water must be used as supply water for washing to eliminate heavy metals such as copper or iron, as they are contaminating impurities.

With these considerations the operating conditions are given in Table 24A and B. In this table, costs are indicated for a plant discharging 460 kg of $NiSO_4$ per day. The net profit is the order of 50 yen per kg of $NiSO_4$ saved. This low profit is mainly due to a high investment cost of the plant, including dialysis stacks, but also resin exchange columns and tanks, and to the cost of pretreatment.

Dialysis of Effluents from Anodic Oxidation of Aluminum

A Donnan dialysis method has been developed for the treatment of effluents discharged in the anodic process for materials in aluminum frame factories. In general, anodic oxidation is carried out in solutions containing about 160 g H_2SO_4 per liter. During the oxidation, the aluminum concentration is controlled and kept below a certain level to maintain the bath composition constant. For this, part of the solution of the anodizing tank is discharged continuously. This waste is then treated in a dialyzer to recover most of the free sulfuric acid, as indicated in Figure 22.

In a practical system, 75% of the free sulfuric acid is recovered. The effluent obtained after recovery of sulfuric acid is generally neutralized and may be efficiently used as a raw material for the preparation of aluminum sulfate. Table 25 gives the results for an operating diffusion dialysis plant. One can see that the loading of the water used in the recovery stream with the sulfuric acid of the diluting stream induces a reduction in volume of about 4%. This volume contraction results from the dissolution of sulfuric acid in water and is always associated with an exotherm. As a result, the temperature of the two output flows from the dialyzer rises. According to local considerations, the temperature of the whole is maintained to a level convenient for the dialysis membranes by means of appropriate methods.

In a process for the fabrication of framing materials with a production rate of 1000 tons of aluminum per month, the quantity of waste solute is approximately 7 m^3/day, and the membrane area required for dialysis is 460 m^2 for dissolved aluminum of 110 kg/day.

Dialysis of Effluents from the Plating Industry

The plating industry has made remarkable progress, mainly in automobile and electronic applications. The applications of plating are very diversified and are

Table 24A Treatment of a Nickel Effluent by DD: Specifications

Effluent treated	
Nickel content discharged	7 kg/day
Concentration	3 g Ni/liter
Recovery rate	90%
Molecular weight	155 g/mole $NiSO_4$
Dialyzer stack (TIII Asahi Glass)	
Membrane spacing	2 mm
Membrane size	0.43×0.90 m × m
Effective area	68%
Number of sheets	100×2 stacks
Flow ratio of waste solution to water	1/4
Cation exchange column	
Cation exchange resin	$300 \text{ I} \times 2$ columns
Adsorption period	1–4 m^3/hr
Regeneration period	0.6 m^3/hr
Unit prices	
Membranes	35,000 Yen/m^2
Electricity	6 Yen/kWh
Nickel	622 Yen/kg $NiSO_4$
Sulfuric acid	15.5 Yen/kg H_2SO_4
Tap water	80 Yen/m^3 H_2O

Source: Data from Ref. 8.

Table 24B Treatment of a Nickel Effluent by DD: Economic Analysis, in yen/kg Nickel Sulfate Saved

Cost of	Abbrev.		
Investment	CUAI		398.92
Pumping	CUFP		10.64
Ion transport water for diffusion dialysis	CUTI		10.43
Pretreatment			
Water for resin columns		7.56	
Regenerating acid		54.93	
Supply of resins		65.22	
Total for pretreatment	CUFT	127.71	127.71
Labor (5% of CUAI)	CUFL		19.95
Total cost	CUI		567.65
Price of nickel purchased	CUVM		622.00
Net profit, in yen/kg $NiSO_4$ saved			54.35

Source: Data from Ref. 8.

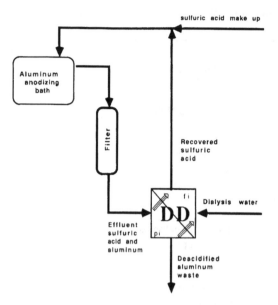

Figure 22   Dialysis of sulfuric acid from aluminum waste.

encountered in multilayer plating of Cu-Ni-Cr in die case products of iron and steel, and of zinc, in anticorrosion plating of zinc, of cadmium, of tin, or in the case of chrome plating for industrial use, or gold and silver plating for IC substrates. Along with these plating processes, there are many kinds of waste acid from metal treatment produced during plating pretreatment. The volume of waste acid varies from small to large, according to the plant where it is used.

As an acid disposal process, neutralization by slaked lime is generally used for small- and medium-scale plating plants, but the cost of sludge disposal is ever

Table 25   Treatment of an Effluent from Anodic Oxidation of Aluminum by Donnan Dialysis

	Diluting stream		Concentrating stream	
Factor	Waste acid	Effluent out	Water in	Recovered acid
Flow rate, liters/hr	275.5	275.5	257.0	243.2
H_2SO_4, g/liter	150	36.7	0	127.4
Aluminum, g/liter	18	16.3	0	1.0

Source: Data from Ref. 49.

increasing, as the pressure of regulations becomes greater. Especially with sulfuric acid waste, the volume of sludge is large.

Accordingly, when using diffusion dialysis in the recovery of sulfuric acid, the cost of sludge disposal is reduced.

Table 26A and B shows specifications and economics of a Japanese diffusion dialysis plant treating 3000 tons of effluent acid per year. Clearly then, diffusion dialysis of acids can save materials and reduce costs. In this case, the turnover of the investments is about 1.5 years.

CONCLUSIONS

Treatment of effluents by dialysis, electrodialysis, or electro-electrodialysis processes improves all the steps in the manufacture of many products. The principal effect of these improvements is the direct or indirect reduction of the consumption of materials. For example, reducing by electrodialysis the amount of water to be evaporated before crystallization also reduces amounts of natural gas or coal used in boilers producing steam for multiple effect evaporators.

These techniques then contribute to saving the earth's resources and energy.

Simply changing the concentration of effluents by electrodialysis, yielding a more concentrated and more dilute solution, increases the different economic and

Table 26A Treatment of Sulfuric Acid Effluent by DD: Specifications

Effluent treated	
Volume	9 m³/day
Concentration of acid	150–250 g H_2SO_4/liter
Concentration of iron	50 g Fe/liter
Recovery rate	80% of acid
Elimination rate of iron	450 kg/day
Dialyzer stack (TSD 50 Tokuyama)	
Membrane spacing	− mm
Dialyzer size	0.6 m × 1.3 m × 1.9 m
Effective area	50 dm²
Number of sheets	1000
Flow ratio of waste solution to water	
Unit prices	
Membranes	− Yen/m²
Electricity	15 Yen/kWh
Sulfuric acid	15 Yen/kg H_2SO_4
Calcium hydroxide	20 Yen/kg $Ca(OH)_2$
Tap water	50 Yen/m³ H_2O

Source: Data from Ref. 54.

Table 26B Treatment of a Sulfuric
Acid Effluent by DD: Economic
Analysis, in yen/m³ Sulfuric Acid
Saved

Cost of	
Utility	244
Concentration agent	29
Ion exchange membranes	282
Others	95
Total for operating costs	650
H_2SO_4 purchase	1937
Cut-down cost of $Ca(OH)_2$	2444
Cut-down of sludging costs	2844
Total for materials recovery	7225

Source: Data from Ref. 54.

material yields and ratios. Furthermore, other improvements directly related to these savings also exist, such as an increase in quality and reduction of pollution.

The use of a change in composition, concentrating on the one hand and diluting on the other, by electrodialysis and related processes, is economically sound. Generally, the payback period of the installation is about 1 year. Dialysis, electrodialysis, and electro-electrodialysis are mainly used to treat aqueous solutions, but some attempts are being made to deal with anhydrous liquid solutions, either organic or inorganic.

As aqueous solutions are the main effluents, if these processes come into general use, the quantities of lost materials should be drastically reduced. These reductions include all secondary industries, such as biological, chemical, electronics, food, metallurgical, nuclear, paper, parachemical (such as paints, perfumes, cosmetics), petroleum, pharmaceutical, wastes, etc., and also many of the primary industries, such as agriculture, arboriculture, breeding, fishing, mining, etc. Moreover, they have been introduced in the medical industries and in everyday life.

Dialysis, electrodialysis, and electro-electrodialysis have now attained industrial maturity. There is a large choice of companies able to provide materials, propose sound projects, build plants or stations, and keep installations in repair. However, the main problems is not only to give potential users, managers, and authorities clear information about these possibilities, but to indicate the operations involved. The few examples given here are intended to increase interest for other applications of these processes.

REFERENCES

1. R. Audinos, Improvement of metal recovery by electrodialysis. *J. Membrane Sci.* 27, 143–154 (1986).

2. *Kirk-Othmer Encyclopedia of Chemical Technology*, John Wiley, New York, 1984.

3. ECE/CHEM 65, Use of disposal of waste water from phosphoric acid and titanium dioxide production, United Nations, New York, 1988.

4. J. C. Bara, *Traitements de surface*, Eyrolles Editeur, Paris, 1988.

5. R. Audinos, Membranes techniques in the chemical industry, Chemie Ingenieur Technik, MS 891/81.

6. K. Spiegler, *Principles of Desalination*, Wiley, New York, 1966.

7. H. Kawate, K. Miaso, and M. Takiguchi, Energy saving in salt manufacture by ion exchange membrane electrodialysis. 6th international symposium on salt, Toronto, Vol. 11, 1983, pp. 471–479.

8. S. Itoi, Electrodialysis of effluent from treatment of metallic surfaces. *Desalination* 28, 1983–205 (1979).

9. E. Korngold, K. Koch, and H. Strathmann, Electrodialysis in advanced waste water treatment. *Desalination* 24, 129–139 (1978)

10. E. Tourneux, Concentration by electrodialysis of rinse effluents (of electrodeposition), Symposium Galvano-Organo—CETIM, Senlis, November 18–19, 1982.

11. R. Audinos and S. Vigneswaran, Electrodialysis. in *Water, Waste Water and Sludge Filtration* (S. Vigneswaran and R. Ben Aim, eds.), CRC Press, 1989.

12. R. Pierrard, Recent progress in electrodialysis. *Ind. Alim. Agric.* 93, 569–581 (1976).

13. S. Hwang and K. Kammermeyer, *Membranes in Separation*, Wiley, New York, 1975.

14. D. Cowan and S. Brown, Effect of turbulence on limiting current electrodialysis cell. *Ind. Eng. Chem.* 51(12), 1445–1448 (1959).

15. T. Nashawaki, Concentration of electrolytes prior to evaporation with an electromembrane process. In *Industrial Processing with Membranes* (R. Lacey and S. Loeb, eds.), Wiley, New York, 1972.

16. R. Audinos, Optimization of solution concentration by electrodialysis. *Chem. Eng. Sci.* 38, 431–439 (1983).

17. N. Lakshminarayanaiah, Permeation of water through cation exchange membranes. *Biophys. J.* 7 511–526 (1967).

18. G. Thau, R. Bloch, and O. Kedem, Water transport in porous and non porous membranes. *Desalination 1*, 128–138 (1966).

19. T. Kressman, P. Stanbridge, F. Tye, and A. Wilson, Transference studies with ion selective membranes. II. Water transference. *Trans. Far. Soc.* 59, 2133–2138 (1963).

20. R. Audinos and R. Zana, Determination des nombres d'hydratation du sulfate de zinc. *J. Chim. Phys.* 78, 183–185 (1981).

21. A. Despic and G. Hill, Electroosmosis in charged membranes. The determination of primary solvatation numbers. *Disc. Far. Soc. 21*, 150–162 (1956).

22. D. Hale and D. McCauley, Structure and properties of heterogeneous cation-exchange membranes. *Trans. Far. Soc.* 57, 135–149 (1961).

23. R. Breslau and I. Miller, A hydrodynamic model of electroosmosis. *Ind. Eng. Fund.* 10(4), 554–565 (1971).

24. T. Brudges and J. Lorimer, Dependence of electroosmosis flow on current density and time. *J. Membrane Sci. 13*, 291–305 (1983).

25. R. Audinos and S. Paci, Water transport during the concentration of waste zinc solution by electrodialysis. *Desalination 67*, 523–545, 1987.

26. R. Robinson and R. Stokes, *Electrolyte Solutions*, Butterworths, Londen, 1959.

27. A. Katchalsky and P. Curran, Non equilibrium thermodynamics in biophysics. Ph.D. dissertation, Harvard University, Cambridge, 1965.

28. L. Onsager, Reciprocal relations in irreversible processes. I and II. *Phys. Rev. 37*, 405–426, *38*, 2265–2279 (1931).

29. S. De Groot, *Thermodynamics of Irreversible Processes*. North Holland, Amsterdam, 1966.

30. R. Audinos, Détermination du courant limite d'électrodialyse par conductivité. *Electrochim. Acta 25*, 405–410 (1980).

31. A. Sonin and M. Isaacson, Optimization of flow design in forced flow electrochemical system with special application to electrodialysis. *Indust. Eng. Chem. Proc. Des. Dev. 13*, 241–248 (1974).

32. W. Katz, The electrochemical reversal process. *Desalination 28*, 31–40 (1979).

33. R. Rautenbach, W. Kopp, R. Hellekes, and G. Van Opbergen, Nitrate reduction of well water by reverse osmosis and electrodialysis. Studies on plant performances and costs. *Desalination 65*, 241–258 (1987).

34. R. Audinos, Aspect éconiomique des procédés à membranes. In *Energétique Industrielle* (P. Le Goff, coordonnateur), Librairie Lavoisier, Paris, tome 3, ChX, 1983, pp. 285–315.

35. A. Kuhn, *Industrial Electrochemical Processes*, Elsevier, Amsterdam, 1971.

36. R. Audinos, Les méthodes de séparation électrocinétiques utilisant des membranes artificielles, Filtra 78, Paris, October 1978, and Methodes et techniques de séparation par l'électricité en milieu condensé fluide, Information Chimie, October 1978.

37. E. Tourneux, Concentration by electrodialysis of rinse effluents of electrodeposition. Colloque Galvano-Organo-Cetim, Senlis, November 18–19, 1982.

38. A. Bonin, Control of the effluent issued from copper electrodeposition after the installation of an electrodialysis apparatus, Report AIF-SRTI, SODETEG, Buc, July 1981.

39. S. B. Tuwiner Concentration of electrolytes from dilute washing. U.S. Patent 3,674,669 (1972), U.S. Patent 3,806,436 (1974).

40. J. Gal, J. Chiapello, M. Perrault, and E. Tourneux, Procédé de régulation d'un électrodialyseur et installation perfectionnée d'électrodialyse. Patents Fr 68 00 072 (1986), U.S. 4,713,156, CD 526 598, Eur 86 202 238 9.

41. S. Shridar, Desalination and recovery of catalyst by electrodialysis. 6th Symposium on Membranes and Membrane Processes, Tübingen, September 4–7, 1989.

42. F. Evangelista, Coupling of membrane processes for waste water treatment. 6th Symposium on Membranes and Membrane Processes, Tübingen, September 4–7, 1989.

43. J. McArdle, J. Piccari, and G. Thornburg, Pickle liquor recovery, 1989 AISE Annual Convention, Pittsburgh, 1989.

44. R. Audinos, S. Blatger, and J. Gimard, Procédé électromembranaire pour régénérer les effluents fluorosiliciques, Filtra 88, Paris, October 18–20, 1988.
45. R. Parent and R. Audinos, Proédé d'obtention d'une solution d'acide fluorhydrique à partir d'une solution d'acide hexafluorosilicique et dispositif pour sa mise en oeuvre. Fr. Patent 88 04 739.
46. Y. Kobuchi, H. Motomura, Y. Noma, and F. Hanada, Application of the ion-exchange membranes to acids recovery by diffusion dialysis, Europe–Japan congress on membranes, Stresa (It), June 18–22, 1984.
47. Aquatech, technical notices.
48. Asahi Chemical Industry Co., technical notice A13, 12/1/1985.
49. Asahi Glass Co., technical notice: Reuse of waste water, 3/8/1987.
50. EIF Ecologie, technical notice 1984.
51. Ionac, technical notices.
52. Ionics, technical notices.
53. SRTI-Corning EIVS, technical notices.
54. Tokuyama Soda Ltd., technical notices, 1985.

Index

Absorber, countercurrent flow, 76
Acetic acid
  recovery by ED/fermentation, 174
  recovery by PV, 169
Acid
  gases
    carbon dioxide, 61
    hydrogen sulfide, 61, 72
  recovery, 289
  treatment by ED, 250
  waste, 296
Activity coefficients, 291
Adsorbent, inorganic, 88
Adsorption, metal, 85
Allen velocity equation, 219
Amino acids, 91
Aluminum, anodic oxidation of, 294
Ammonium bifluoride, 280
Applications, 195
  range of, 1, 5, 10
  membrane processes, 4, 7, 10
  separation, 197

Azeotrope
  ethanol/water, 178
  isopropanol/water, 154

Biochemicals, produced by
    fermentation, 169
Bioconversion, 99
Biomass, 99
Bioreactor
  electrodialysis, 178
  membrane, 100
  PV with PTFE/SR and PTMSP
    membranes, 123
  PV, PTMSP, PTFE/SR, 122
Biotechnology, 10
Bond
  pi, 20
  sigma, 20
  tenacity, 59
Bonding model
  Dewar-Chatt, 62, 63
  Donor-acceptor, 62

[Bonding model]
   equilibrium, 69
Bones, and gelatin, 89, 94
Breakthrough, 95
Butadiene, recovery of, 19
Butanol, membrane recovery from
            ABE fermentation broth, 163

Calorimetry, of cuprous complexes in
            solutions, 30
Capacities
   maximum, 86, 89
   volumetric, of copper nitrate and
            CuTFA, 36
Carbon monoxide, recovery by cuprous
            salt solutions, 51
Catalyst, recovery by ED, 271
Catholythe compartment, 285
Chloranate, 21
Chromic acid, 284, 286
Chromic bath, 286
Chromium
   cation, Cr(III), 285
   cation, Cr(VI), 285
      removal, 197
   recovery by EED, 284
Column
   design, 214
   extraction, 213, 225
   flooding, 213
   flow diagram, 216
   glass, 214
   hydrodynamics, 213
   ion exchange resin, 291
   PSE, 213, 222
   pulsed sieve-plate, 213
   vent, 77
Complex
   metal-olefin, 19
   olefin-Ag(I), -Cu(I), -Hg(II), 19
   olefin-Pt(II), 19
   platinum, 62

[Complex]
   platinum(II)-chloride-ethylene, 19
Complexation
   in CuTFA solutions, 29
   heat of, 69
   irreversible, 63
   π-bond, 62
Complexing
   by solid CuTFA, 26
   cuprous-olefin with ionic cuprous
            salts, 26
Concentration
   by electrodialysis, theory, 240
   polarization layers, 291
   safe operating, 76
Concentration profiles, 220
Constriction factor, 219
Contactors
   column, 94
   countercurrent, 96
   fixed/fluidized-bed, 94-97
   stirred-tank, 93
Copper, recovery by ED, 262
Copper nitrate
   stability in propionitrile solutions, 55
   thermal stability, 54
Cost, of energy, 59, 79
Cuprammonium acetate system, 20
Cuprammonium process, 19
Cupric ion, 21
Cuprous
   acetate, 20
   halides, 19
   ion, 19, 21
Cuprous ion-olefin bond, 20
Cuprous salts
   anhydride and Cu_2O, 22
   cuprous-group III halide reaction, 23
   double decomposition, 23
   dry, 22
   ionic, preparation, 22
   in nonaqueous solutions, 23

[Cuprous salts]
  reductive displacement
    by copper, 23
    by silver, 23
  solubilities in nonaqueous solvents,
    24, 25
  of strong acids, 20
Cuprous tetrachloroalanate, 21
Cuprous tetrafluoroborate, 21
Cuprous trifluoroacetate, 21
CuTFA, thermal stability, 54
Cyanide, 197
  baths, treatment of effluents by ED,
    265
  removal, 197

Demulsification, 196
Detonation, 63
Dialysis
  acid, 289, 291
  coefficient, 291
  diffusion, 291, 297
  Donnan diffusion, 232, 286, 289
  molecular, 288
  ordinary, 286
  stack, 289, 291
Diffusion, advancing front, 196
Diolefins, separation by cuprous
    complexing, 50
Disproportionation
  equilibrium, 21
  reaction, 20
Distillation
  advanced, 104
  azeotropic, 117
  conventional, separation of olefins,
    33
  cryogenic, 59
  extractive, 104
    propylene/propane separations, 38
    separation of olefins, 33

[Distillation]
  vapor recompression, 104
Distribution
  coefficient, 86, 88
  regional, 10, 13
  market, 12
Driving force, mean logarithmic,
    291
Drop
  digitizer, 217
  mean size, 218, 221
    correlation, 225
  size distribution, 218
Drops
  hydrodynamic equation, 226

Economic analysis, 79
  annual costs, 254
  costs of products treated, 255
  of ED processes, 252
  investment costs, 253
Effluents, aqueous solutions, 229, 298
Electric current density, 242
Electrochemical potential, 246, 247
Electrodeposition bath, 284
Electrodialysis, 100, 101
  applications, 232
  batch, 237
  continuous, 237
  operating characteristics, 238
  operating conditions, 244
  operating modes, 237
  operating restrictions, 238
  operation location, 239
  process description, 236
  reversal, 251
Electrodialyzer
  three-stage, 268
  tortuous path, 246, 248
Electro-electrodialysis (EED), principle
    of, 232, 283

Electrolyte
  flux, 247
  strong, 65
Electromembrane processes, 232
Electroosmotic transport number, 244
Electrowinning of metals, 256
Emulsion
  breaking, 208
  formulation, 196
Encapsulation, 207
Energy, cost of, 59, 79
Equilibria
  competitive
  CuTFA-olefincomplexation
  Donnan, 289
  phase, 66
  vapor-liquid, 59
Equilibrium constants
  calorimetric determination, 54
  of propylene in CuTFA solutions, 37
Equipment, manufacture of, 15
Ethanol concentration
  by CCRO/MSFD/PV process, 150
  from low-cellulose lignocellulosic
    feedstocks, 148
  by hybrid RO/distillation
  by multistage flash/azeotropic
    distillation, 148
  by RO, continuous membrane
    column, CCRO, 149
Ethanol dehydration
  by conventional distillation, 179
  energy requirements by distillation,
    178
  by hybrid distillation/PV process,
    106
  PV plant, 116
  by reverse osmosis, 150
Ethanol/organic separations
  by hybrid PV/liquid-liquid
    extraction, 134
  by hybrid RO/extraction process, 150

[Ethanol/organic separations]
  by reverse osmosis, 150
Ethanol recovery
  energy requirements, 105
  by hybrid distillation/VPe process,
    139
  process selection based on
    concentration, 175
Ethyl acetate, recovery by PV, 169
Extraction, liquid, 104

Facilitation
  transported, 64, 76
  types 1 and 2, 196
Feedstocks, lignocellulosic, 104
Fermentation
  continuous ABE, 154
    productivity, 156
  continuous IBE, 154, 157
    productivity, 156
  extractive, 104
Ferrous sulfate, as a pollutant, 230
Filter press, assembly, 289
Flow regimes, emulsion, 219
Fluid
  drilling, 207
  shear-thickening, 207
Flux
  definition, 102
  permeation, 288
Fusel oils, recovery by PV, 169
Future improvements, 208

Gels
  alginate, 91
  matrix, 93
Glass-etching solution, recovery, 279
Glass industry
  application of ED, 268
  demineralization of rinse water, 268
Gold, recovery from wastewater, 197

Gradient, interfacial tension, 221

Heat pump, 78
Hexafluorosilicic acid, 280
Holdup, dispersed phase, 218
Hydrofluoric/nitric acid mixture, 278
Hydrometallurgical processes,
        application of ED, 256
Hypochlorite, 283

Internal phase, 199
Ion exchange, 91
    sites, 92
Ion pump, 199
Iron, cation, Fe(III), 285
Isoprene complexing, 20
Isopropanol
    dehydration by hybrid process, 160
    dehydration by VPe, 161
Isotherms
    adsorption, 87, 89
    BET, 88
    Freudlich, 88
    Langmuir, 88
    propylene and propane from
        CuTFAlene, 35

Lactic acid, recovery by ED bioreactor,
    174
Ligands, bridging, 20
Liquid membrane
    emulsion, 195, 198
    supported, 198
Loading
    equilibrium, 89
    maximum, 88

Mass transfer
    coefficient, 214, 221, 226
        experimental/predicted, 224
        PSE column, 217, 221
    Handlos-Baron model, 221, 223

[Mass transfer]
    HTU, 214
    individualism, 221
    Kronig-Brink model, 221, 223
    PSE column, 214
    single-drop model, 221, 223
McCabe-Thiele, 103
Mechanism
    adsorption, 90
    desorption, 90, 92
    interaction, 85
    single-drop movement, 219
Medical devices, 9
Membrane
    anion exchange, 236
        selectivity, 290
        bipolar, 209
        water separation into ions, 276
    carboxymethylcellulose (CMC), 106
    cation exchange, 236
    cellulose acetate (CA), 106
    cellulose acetate RO, 144
    composite, 4
    cost, 2
    definition of, 3
    electrohydrolysis, 276
    ethanol-selective, 119-121
    fouling, 8, 16
    GFT polyvinyl alcohol (PVA), 106
    hexafluoroethaneallylamine, 122
    ion exchange, 4, 101, 277
    limitations, 2
    liquid, 195, 198
    market, 10
    microfiltration, 4
    modules, 4, 7, 10, 13
    monopolar, 277
    Nafion, 106
    PAA/GPC, 106
    performance, 2, 102
    polyacrylic (PAA), 106
    polyacrylonitrile (PAN), 106

[Membrane]
polydimethylsiloxane, 120
polyhydroxymethylene-
    cofluoroolefin, 106
polypropylene, 122
polytetrafluoroethylene, 122
polyvinyl fluoride (PVD), 106
polyvinylidene fluoride (PVDF), 106
    radiation-cured, 106
potentials, 2
process, 64
properties, 5
PTFE, hollow-fiber modules for
    ethanol stripping, 122
PTMSP, 122
PVF/4-vinylpyrrolidone(betaine),
    106
selectivity, 16
semipermeable, 286
structure, 3
ultrafiltration, 4
unit operations, 100
water-selective, 103, 106
Membrane extraction
in biochemicals production, 175
recovery of ABE, 163
Metal deposition, in metals production,
    231
Metal fluoride complex, 278
Microorganisms, 85
Milk byproducts, treatment by ED, 272
Model(ing), 65
backflow, 219
movement, 223
Molecular orbital, 20

Nernst-Einstein equation, 241
Nickel
dialysis of effluents from
    galvanization, 291
recovery from by galvanization by
    ED, 259

Nickel salts
chloride, 262
sulfate, 262
Nickel sulfate, 262
solution, 294
Nutrients, necessary, 91

Olefin/paraffin, separation, 19
Operating regimes
emulsion, 213
flooding, 213
mixed-settler, 213
unstable, 213
Optimization, of existing systems, 4
Osmosis, 204
Osmotic pressure, 246
Oxidation
in situ, 75
state, 91

Parameter estimation, 219
Peripheral components, 14
Permeability, definition of, 102
Permeate, 100
Perstaction
membrane extraction, 100-101
product recovery during continuous
    fermentation, 151
Pervaporation
biochemicals production,169
concentration, organic-selective
    membranes, 154
continuous membrane column, 117
continuous two-column, 117
dehydration, water-selective
    membranes, 152
ethanol concentration, 119
ethanol/organic separations, 134
ethanol/water data, 106
    polyacrylic acid membranes, 106
    polysulfone membranes, 106
PV, 100-101

Pervaporator, two-column, 132-133
Phase
    aqueous, 214
    continuous, 214
    dispersed, 214
    internal, 199
    organic, 214
Phase diagrams, ternary, 56
Phenol, removal of, 197, 205
Plating
    multilayer Cu-Ni-Cr, 294
    zinc, cadmium, tin, chrome, gold,
        silver, 294
Polarization parameter, 242
Pollutants
    radioactive, 86
    removal of, 85, 97
Pollution, 229
Prerefining, 71
    steps, 73
Pretreatment
    chemical and physical, 86
    in metals production, 231
Price structure, 14
Process
    ESEP, 37
    olefin/paraffin separation, 33
    risk, 79
Profitability, 79
Propionitrile, 21
Propylene glycol alginate, 89
Pump
    heat, 78
    ion, 199
Pyrolisis, 60

Rate, optimum, 88
Reagent, 199
Recovery
    acetone, 151
    butadiene, 19
    butanol, 151, 163

[Recovery]
    carbon monoxide, 51
    catalyst, 271
    isopropanol, 151
    quantitative, 89
    styrene, by dual-solvent liquid-liquid
        extraction, 49
Recycling mode, 93
Reduction, to silver, 63
Rejection
    coefficient, 102
    definition, 102
Research and development, 14, 17
Retentate, 100
Reverse osmosis
    biochemicals production, 174
    ethanol concentration, 141
    products concentration, 161

Sales
    annual, 12
    in chemical industries, 11
    of medical devices, 10
    of membrane industry, 9
    of synthetic membranes, 1
    volume, 14
    worldwide, 13
Scale, formation of, 250
Selectivity, 39
    definition, 102
Separation
    detergent range olefins from
        paraffins, 46
    high molecular weight olefins from
        paraffins, 38
    low molecular weight olefins from
        paraffins, 33
    product/organic, 169
    styrene/ethylbenzene, 44
Separation factor, definition of, 102
Sherwood number, 221-222
Silica precipitate, 282

Sites
   binding, 91
   ion exchange, 92
Sodium sulfate, as a pollutant, 230
Solubilities, 27, 76
Solubilizer, 201
Solution
   detoxification, 91
   pH, 87
Sorption, 104
Stability
   chemical and physical, 8
   of chlorine, 16
   of solution, 74, 81
Steel pickle liquor, recovery, 278
Steric hindrance, 63
Stripping, CO_2
Structure
   asymmetric, 4
   microporous, 4
Styrene, recovery, 49
Sulfuric acid, 294
   pollutant, 230
Surface modification, in metals
            production, 231
Sweep gas, 101
Swell, 204
System
   oxylene/acetone/water, 218
   water/succinic acid/n-butanol, 221

Temperature, subambient, 59
Time
   contact, 88, 96
   exposure, 86
   residence, 202
   retention, 95

Tissue, animal, 89
Transport
   active, 4
   facilitated, 64, 76
   modes, 3

Ultrafiltration (UF), 100
Unloading, 88

Vapor permeation
   ethanol concentration, 141
   ethanol dehydration, 135
   ethanol/water data, 139
   products dehydration, 161
   VPe, 100-101
Velocity, slip, 219, 221
Venting, of impurities, 76
Vitreous silica, 280

Wastes, nonradioactive from nuclear
            plants, 231
Wastewater, 205
   treatment by ED, 272
Water desalination, 234
Water treatment, 8
Whey, treatment by ED, 272

Xylose, ethanol production from, 104

Zinc metallurgy
   application of ED, 257
   operating modes, 258
   recovery, 197, 200, 204
   single stage ED, 258
   two-stage ED, 259
Zinc sulfate, concentration of solutions,
            246

9 780367 450274